The Case against Perfection

The Case against Perfection

ETHICS IN THE AGE OF GENETIC ENGINEERING

Michael J. Sandel

THE BELKNAP PRESS OF
HARVARD UNIVERSITY PRESS
Cambridge, Massachusetts, and London, England

Copyright © 2007 by Michael J. Sandel

PRINTED IN THE UNITED STATES OF AMERICA

Library of Congress Cataloging-in-Publication Data

Sandel, Michael J.
The case against perfection : ethics in the age of genetic
engineering / Michael J. Sandel.
p. cm.
Includes bibliographical references and index.
ISBN-13: 978-0-674-01927-0 (alk. paper)
ISBN-10: 0-674-01927-X (alk. paper)
1. Genetic engineering—Moral and ethical aspects.   I. Title.
QH438.7.S2634   2007
174′.957—dc22   2006039327

For Adam and Aaron

Acknowledgments

My interest in ethics and biotechnology was prompted by an unexpected invitation, in late 2001, to serve on the newly formed President's Council on Bioethics. Although I am not a professional bioethicist, I was intrigued by the prospect of thinking my way through controversies over stem cell research, cloning, and genetic engineering in the company of a distinguished group of scientists, philosophers, theologians, physicians, legal scholars, and public policy experts. I found the discussions enormously stimulating and intellectually intense, so much so that I decided to pursue some of the topics in my teaching and writing. Leon Kass, who chaired the council during the four years I served, was largely responsible for the high level of the discussions. Although he and I have strong philosophical and political differences, I admire Leon's unerring eye for important questions and am grateful to him for having embroiled the council, and me, in far-reaching bioethical inquiries the likes of which few governmental bodies undertake.

One of the questions that most intrigued me concerned the ethics of genetic enhancement. I wrote a short discussion paper on the subject for the council, and, with the encouragement of Cullen Murphy, developed it into an essay for the *Atlantic Monthly* in 2004. Cullen is a writer's dream editor—a smart, sympathetic critic with a keen moral sensibility and exquisite editorial judgment. I am indebted to Cullen for suggesting the title of this book, and for nurturing the essay of the same title that first appeared in the pages of his magazine. I am also grateful to Corby Kummer, who helped edit the essay from which this book was born.

For the past several years I have had the privilege of exploring the themes of this book with Harvard undergraduates, graduate students, and law students in seminars I have taught on ethics and biotechnology. In 2006 I teamed up with my colleague and friend Douglas Melton to teach a new undergraduate course, Ethics, Biotechnology, and the Future of Human Nature. More than a distinguished biologist and stem cell pioneer, Doug has the philosopher's knack for asking seemingly innocent questions that go to the heart of the matter. It has been a great pleasure to explore these questions in his company.

I am grateful for having had the opportunity to try out various of the arguments presented in this book in the Moffett Lecture at Princeton University; the Geller Lecture at NYU School of Medicine; the Dasan Memorial Lecture in Seoul, South Korea; a public lecture at an international conference in Berlin organized by the Deutsches Referenzzentrum für Ethik in den Biowissenschaften (DRZE); a public lecture at the Collège de France, Paris; and a bioethics colloquium cosponsored by the National Institutes of Health, Johns Hopkins University, and Georgetown University. I learned a great deal from the comments and criticisms offered by participants in those occasions. I am also grateful for the support of the Harvard Law School summer research program, and the Carnegie Scholars program of the Carnegie Corporation, which graciously allowed me this intellectual detour along the way to a future (and not wholly unrelated) project on the moral limits of markets.

I would like also to record my thanks to Michael Aronson, my editor at Harvard University Press, who has guided this book to completion with exemplary patience and care, and to Julie Hagen for her fine copy editing. Finally, I am indebted above all to my wife, Kiku Adatto, whose intellectual and spiritual sensibilities improved this book and me. I dedicate the book to our sons, Adam and Aaron, who are perfect just as they are.

Contents

The Case against Perfection

I

The Ethics of Enhancement

A FEW YEARS AGO, a couple decided they wanted to have a child, preferably a deaf one. Both partners were deaf, and proudly so. Like others in the deaf-pride community, Sharon Duchesneau and Candy McCullough considered deafness a cultural identity, not a disability to be cured. "Being deaf is just a way of life," said Duchesneau. "We feel whole as deaf people and we want to share the wonderful aspects of our deaf community—a sense of belonging and connectedness—with children. We truly feel we live rich lives as deaf people."[1]

In hopes of conceiving a deaf child, they sought out a sperm donor with five generations of deafness in his family. And they succeeded. Their son Gauvin was born deaf.

The new parents were surprised when their story, which was reported in the *Washington Post*, brought widespread condemnation. Most of the outrage focused on the charge that they had deliberately inflicted a disability on their child. Duchesneau and McCullough (who are lesbian partners) denied that deafness is a disability and argued that they had simply wanted a child like themselves. "We do not view what we did as very different from what many straight couples do when they have children," said Duchesneau.[2]

Is it wrong to make a child deaf by design? If so, what makes it wrong—the deafness or the design? Suppose, for the sake of argument, that deafness is not a disability but a distinctive identity. Is there still something wrong with the idea of parents picking and choosing the kind of child they will have? Or do parents do that all the time, in their choice of mate and, these days, in their use of new reproductive technologies?

Not long before the controversy over the deaf child, an ad appeared in the *Harvard Crimson* and other Ivy League student newspapers. An infertile couple was seeking an egg donor, but not just any egg donor. She had to be five feet, ten inches tall, athletic, without major family medical problems, and to have a combined SAT score of 1400 or above. In exchange for an egg from a donor meeting this description, the ad offered payment of $50,000.[3]

Perhaps the parents who offered the hefty sum for a premium egg simply wanted a child who resembled them. Or perhaps they were hoping to trade up, trying for a child who would be taller or smarter than they. Whatever the case, their extraordinary offer did not prompt the public outcry that met the parents who wanted a deaf child. No one objected that height, intelligence, and athletic prowess are disabilities that children should be spared. And yet something about the ad leaves a lingering moral qualm. Even if no harm is involved, isn't there something troubling about parents ordering up a child with certain genetic traits?

Some defend the attempt to conceive a deaf child, or one who will have high SAT scores, as similar to natural procreation in one crucial respect: whatever these parents did to increase the odds, they were not guaranteed the outcome they sought. Both attempts were still subject to the vagaries of the genetic lottery. This defense raises an intriguing question. Why does some element of unpredictability seem to make a moral difference? Suppose biotechnology could remove the uncertainty and allow us to design the genetic traits of our children?

While pondering this question, put aside children for a moment and consider pets. About a year after the furor over the deliberately deaf child, a Texas woman named Julie (she declined to give her last name) was mourning the death of her beloved cat Nicky. "He was very beautiful," Julie said. "He was exceptionally intelligent. He knew eleven commands." She had read of a company in California that offered a cat cloning service—Genetic Savings & Clone. In 2001 the company had succeeded in creating the first cloned cat (named CC, for Carbon Copy). Julie sent the company a genetic sample of Nicky, along with the required fee of $50,000. A few months later, to her great delight, she received Little Nicky, a genetically identical cat. "He is identical," Julie proclaimed. "I have not been able to see one difference."[4]

The company's Web site has since announced a price reduction for cat cloning, which now costs a mere $32,000. If the price still seems steep, it comes with a money-back guarantee: "If you feel that your kitten doesn't sufficiently resemble the genetic donor, we'll refund your money in full with no questions asked." Meanwhile, the company's scientists are working to develop a new product line—cloned dogs. Since dogs are harder to clone than cats, the company plans to charge $100,000 or more.[5]

Many people find something odd about the commercial cloning of cats and dogs. Some complain that, with thousands of strays in need of good homes, it is unconscionable to spend a small fortune to create a custom-made pet. Others worry about the number of animals lost during pregnancy in the attempt to create a successful clone. But suppose these problems could be overcome. Would the cloning of cats and dogs still give us pause? What about the cloning of human beings?

Articulating Our Unease

Breakthroughs in genetics present us with a promise and a predicament. The promise is that we may soon be able to treat and prevent a host of debilitating diseases. The predicament is that our newfound genetic knowledge may also enable us to manipulate our own nature—to enhance our muscles, memories, and moods; to choose the sex, height, and other genetic traits of our children; to improve our physical and cognitive capacities; to make ourselves "better than well."[6] Most people find at least some forms of genetic engineering disquieting. But it is not easy to articulate the source of our unease. The familiar terms of moral and political discourse make it difficult to say what is wrong with reengineering our nature.

Consider again the question of cloning. The birth of Dolly the cloned sheep in 1997 brought a torrent of worry about the prospect of cloned human beings. There are good medical reasons to worry. Most scientists agree that cloning is unsafe and likely to produce offspring with serious abnormalities and birth defects. (Dolly died a premature death.) But suppose cloning technology improves to the point where the risks are no greater than with natural pregnancy. Would human cloning still be objectionable? What exactly is wrong with creating a child who is a genetic twin of his or her parent, or of an older sibling who has tragically died, or, for that matter, of an admired scientist, sports star, or celebrity?

Some say cloning is wrong because it violates the child's right to autonomy. By choosing in advance the genetic makeup of the child, the parents consign her to a life in the shadow of someone who has gone before, and so deprive the child of her right to an open future. The autonomy objection can be raised not only against cloning but also against any form of bioengineering that allows parents to choose their child's genetic characteristics. According to this objection, the problem with genetic engineering is that "designer children" are not fully free; even favorable genetic enhancements (for musical talent, say, or athletic prowess) would point children toward particular life choices, impairing their autonomy and violating their right to choose their life plan for themselves.

At first glance, the autonomy argument seems to capture what is troubling about human cloning and other forms of genetic engineering. But it is not persuasive, for two reasons. First, it wrongly implies that, absent a designing parent, children are free to choose their physical characteristics for themselves. But none of us chooses our own genetic inheritance. The alternative to a cloned or genetically enhanced child is not one whose future is unbiased and unbound by particular talents, but a child at the mercy of the genetic lottery.

Second, even if a concern for autonomy explains some of our worries about made-to-order children, it cannot explain our moral hesitation about people who seek genetic enhancements for themselves. Not all genetic interventions are passed down the generations. Gene therapy on nonreproductive (or somatic) cells, such as muscle cells or brain cells, works by repairing or replacing defective genes. The moral quandary arises when people use such therapy not to cure a disease but to reach beyond health, to enhance their physical or cognitive capacities, to lift themselves above the norm.

This moral quandary has nothing to do with impairing autonomy. Only germline genetic interventions, which target eggs, sperm, or embryos, affect subsequent generations. An athlete who genetically enhances his muscles does not confer on his progeny his added speed and strength; he cannot be charged with foisting talents on his children that may push them toward an athletic career. And yet there is still something unsettling about the prospect of genetically altered athletes.

Like cosmetic surgery, genetic enhancement employs medical means for nonmedical ends — ends unrelated to curing or preventing disease, repairing injury, or restoring health. But unlike cosmetic surgery, genetic enhancement is not merely cosmetic. It is more than skin deep. Even somatic enhancements, which would not reach our children and grandchildren, raise hard moral questions. If we are ambivalent about plastic surgery and Botox injections for sagging chins and furrowed brows, we are all the more troubled by genetic engineering for stronger bodies, sharper memories, greater intelligence, and happier moods. The question is whether we are right to be troubled—and if so, on what grounds?

When science moves faster than moral understanding, as it does today, men and women struggle to articulate their unease. In liberal societies, they reach first for the language of autonomy, fairness, and individual rights. But this part of our moral vocabulary does not equip us to address the hardest questions posed by cloning, designer children, and genetic engineering. That is why the genomic revolution has induced a kind of moral vertigo. To grapple with the ethics of enhancement, we need to confront questions largely lost from view in the modern world—questions about the moral status of nature, and about the proper stance of human beings toward the given world. Since these questions verge on theology, modern philosophers and political theorists tend to shrink from them. But our new powers of biotechnology make them unavoidable.

GENETIC ENGINEERING

To see how this is so, consider four examples of bioengineering already on the horizon: muscle enhancement, memory enhancement, height enhancement, and sex selection. In each case, what began as an attempt to treat a disease or prevent a genetic disorder now beckons as an instrument of improvement and consumer choice.

Muscles

Everyone would welcome a gene therapy to alleviate muscular dystrophy and to reverse the debilitating muscle loss that comes with old age. But what if the same therapy were used to produce genetically altered athletes? Researchers have developed a synthetic gene that, when injected into the muscle cells of mice, makes muscles grow and prevents them from deteriorating with age. The success bodes well for human applications. Dr. H. Lee Sweeney, who leads the research, hopes his discovery will cure the immobility that afflicts the elderly.

But Dr. Sweeney's bulked-up mice have already attracted the attention of athletes seeking a competitive edge.[7] The gene not only repairs injured muscles but also strengthens healthy ones. Although the therapy is not yet approved for human use, the prospect of genetically enhanced weight lifters, home-run sluggers, linebackers, and sprinters is easy to imagine. The widespread use of steroids and other performance-enhancing drugs in professional sports suggests that many athletes will be eager to avail themselves of genetic enhancement. The International Olympic Committee has already begun to worry about the fact that, unlike drugs, altered genes cannot be detected in urine or blood tests.[8]

The prospect of genetically altered athletes of fers a good illustration of the ethical quandaries surrounding enhancement. Should the IOC and professional sports leagues ban genetically enhanced athletes, and if so, on what grounds? The two most obvious reasons for banning drugs in sports are safety and fairness: Steroids have harmful side effects, and to allow some to boost their performance by incurring serious health risks would put their competitors at an unfair disadvantage. But suppose, for the sake of argument, that muscle-enhancing gene therapy turned out to be safe, or at least no riskier than a rigorous weight-training regime. Would there still be a reason to ban its use in sports? There is something unsettling about the specter of genetically altered athletes lifting SUVs or hitting 650-foot home runs or running a three-minute mile. But what exactly is troubling about these scenarios? Is it simply that we find such superhuman spectacles too bizarre to contemplate, or does our unease point to something of ethical significance?

The distinction between curing and improving seems to make a moral difference, but it is not obvious what the difference consists in. Consider: If it is all right for an injured athlete to repair a muscle tear with the help of genetic therapy, why is it wrong for him to extend the therapy to improve the muscle, and then to return to the lineup better than before? It might be argued that a genetically enhanced athlete would have an unfair advantage over his unenhanced competitors. But the fairness argument against enhancement has a fatal flaw. It has always been the case that some athletes are better endowed, genetically, than others. And yet we do not consider the natural inequality of genetic endowments to undermine the fairness of competitive sports. From the standpoint of fairness, enhanced genetic differences are no worse than natural ones. Moreover, assuming they are safe, genetic enhancements could be made available to all. If genetic enhancement in sports is morally objectionable, it must be for reasons other than fairness.

Memory

Genetic enhancement is possible for brains as well as brawn. In the mid-1990s scientists managed to manipulate a memory-linked gene in fruit flies, creating flies with photographic memories. More recently researchers produced smart mice by inserting extra copies of a memory-related gene into mouse embryos. The altered mice learn more quickly and remember things longer than normal mice. For example, they are better able to recognize objects they have seen before, and to remember that a certain sound leads to an electric shock. The gene the scientists tweaked in mouse embryos is present in human beings as well, and becomes less active as people age. The extra copies installed in the mice were programmed to remain active even in old age, and the improvement was passed on to their offspring.[9]

Of course human memory is more complicated

than recalling simple associations. But biotech companies with names like Memory Pharmaceuticals are in hot pursuit of memory-enhancing drugs, or "cognition enhancers," for human beings. One obvious market for such drugs consists of those who suffer from serious memory disorders, such as Alzheimer's and dementia. But the companies also have their sights on a bigger market: the 76 million baby boomers over fifty who are beginning to encounter the natural memory loss that comes with age.[10] A drug that reversed age-related memory loss would be a bonanza for the pharmaceuticals industry, a "Viagra for the brain."

Such use would straddle the distinction between remedy and enhancement. Unlike a treatment for Alzheimer's, it would cure no disease. But insofar as it restored capacities a person once possessed, it would have a remedial aspect. It could also have purely nonmedical uses: for example, by a lawyer cramming to memorize facts for an upcoming trial, or by a business executive eager to learn Mandarin on the eve of his departure for Shanghai.

It might be argued, against the project of memory enhancement, that there are some things we would rather forget. For the drug companies, however, the desire to forget represents not an objection to the memory business but another market segment. Those who want to blunt the impact of traumatic or painful memories may soon be able to take a drug that prevents horrific events from being etched too vividly in memory. Victims of a sexual assault, soldiers exposed to the carnage of war, or rescue workers forced to face the aftermath of a terrorist attack would be able to take a memory-suppressing drug to dull the trauma that might otherwise plague them for a lifetime. If the use of such drugs became widely accepted, they might one day be administered routinely in emergency rooms and military field hospitals.[11]

Some who worry about the ethics of cognitive enhancement point to the danger of creating two classes of human beings—those with access to enhancement technologies, and those who must make do with an unaltered memory that fades with age. And if the enhancements can be passed down the generations, the two classes may eventually become subspecies of human beings—the enhanced and the merely natural. But the worry about access begs the question of the moral status of enhancement itself. Is the scenario troubling because the unenhanced poor are denied the benefits of bioengineering, or because the enhanced affluent are somehow dehumanized? As with muscles, so with memory: The fundamental question is not how to assure equal access to enhancement but whether we should aspire to it. Should we devote our biotechnological ingenuity to curing disease and restoring the injured to health, or should we also seek to improve our lot by reengineering our bodies and minds?

Height

Pediatricians already struggle with the ethics of enhancement when confronted by parents who want to make their children taller. Since the 1980s, human growth hormone has been approved for children with a hormone deficiency that makes them much shorter than average.[12] But the treatment also increases the height of healthy children. Some parents of healthy children who are unhappy with their stature (typically boys) ask for the hormone treatments on the grounds that it should not matter whether a child is short because of a hormone deficiency or because his parents happen to be short. Whatever the cause, the social consequences of shortness are the same in both cases.

In the face of this argument, some doctors began prescribing hormone treatments for children whose short stature was unrelated to any medical problem. By 1996 such "off-label" use accounted for 40 percent of human growth hormone prescriptions.[13] Although it is not illegal to prescribe drugs for purposes the Food and Drug Administration (FDA) has not approved, the pharmaceutical companies cannot promote such use. Seeking to expand its market, one company, Eli Lilly, recently persuaded the FDA to approve its human growth hormone for healthy children whose projected adult height is in the bottom first percentile—under five feet, three inches for boys; four feet, eleven inches for girls.[14] This small concession raises a large question about the ethics of enhancement: If hormone treatments need not be limited to those with hormone deficiencies, why should they be available only to very short children? Why shouldn't all shorter-than-average children be able to seek treatment? And what about a child of average height who wants to be taller so he can make the basketball team?

Critics call the elective use of human growth hormone "cosmetic endocrinology." Health insurance is unlikely to cover it, and the treatments are expensive. Injections are administered up to six times a week, for two to five years, at an annual cost of about $20,000—all for a potential height gain of two or three inches.[15] Some oppose height enhancement on the grounds that it is collectively self-defeating; as some become taller, others will become shorter relative to the norm. Except in Lake Wobegon, every child cannot be above average in height. As the unenhanced begin to feel shorter, they too might seek treatment, leading to a hormonal arms race that will leave everyone worse off, especially those who cannot afford to buy their way up from shortness.

But the arms-race objection is not decisive on its own. Like the fairness objection to bioengineered muscles and memory, it leaves unexamined the attitudes and dispositions that prompt the drive for enhancement. If we were bothered only by the injustice of adding shortness to the problems of the poor, we could remedy that unfairness by providing publicly subsidized height enhancement. As for the collective-action problem, the innocent bystanders who suffer relative height deprivation could be financially compensated by a tax imposed on those who buy their way to greater height. The real question is whether we want to live in a society where parents feel compelled to spend a fortune to make perfectly healthy kids a few inches taller.

Sex Selection

Perhaps the most alluring nonmedical use of bioengineering is sex selection. For centuries parents have been trying to choose the sex of their children. Aristotle advised men who wanted a boy to tie off their left testicle before intercourse. The Talmud teaches that men who restrain themselves and allow their wives to achieve sexual climax first will be blessed with a son. Other recommended methods have involved timing conception in relation to ovulation, or to the phases of the moon. Today, biotech succeeds where folk remedies failed.[16]

One technique for sex selection arose with prenatal tests using amniocentesis and ultrasound. These medical technologies were developed to detect genetic abnormalities, such as spina bifida and Down syndrome. But they can also reveal the sex of a fetus, allowing for the abortion of a fetus of the undesired sex. Even among those who favor abortion rights, few advocate abortion simply because the mother (or father) does not want a girl. But in societies with powerful cultural preferences for boys, ultrasound sex determination followed by the abortion of female fetuses has become a familiar practice. In India, the number of girls per 1,000 boys has dropped from 962 to 927 in the past two decades. India has banned the use of prenatal diagnosis for sex selection, but the law is rarely enforced. Itinerant radiologists with portable ultrasound machines travel from village to village, plying their trade. One Bombay clinic reported that, of 8,000 abortions it performed, all but one were for purposes of sex selection.[17]

But sex selection need not involve abortion. For couples undergoing in vitro fertilization (IVF), it is possible to choose the sex of the child before the fertilized egg is implanted in the womb. The procedure, known as preimplantation genetic diagnosis (PGD), works like this: Several eggs are fertilized in a petri dish and allowed to grow to the eight-cell stage (for about three days). At that point, the early embryos are tested to determine their sex. Those of the desired sex are implanted; the others are typically discarded. Although few couples are likely to undergo the difficulty and expense of IVF simply to choose the sex of their child, embryo screening is a highly reliable means of sex selection. And as our genetic knowledge increases, it may be possible to use PGD to cull embryos carrying other undesired genetic traits, such as those associated with obesity, height, and skin color. The 1997 science fiction movie *Gattaca* depicts a future in which parents routinely screen embryos for sex, height, immunity to disease, and even IQ. There is something troubling about the *Gattaca* scenario, but it is not easy to identify what exactly is wrong with screening embryos to choose the sex of our children.

One line of objection draws on arguments familiar from the abortion debate. Those who believe that an embryo is a person reject embryo screening on the same grounds that they reject abortion. If an eight-cell embryo growing in a petri dish is morally equivalent to a fully developed human being, then discarding it is no better than aborting a fetus, and both practices are equivalent to infanticide. Whatever its merits, however, this "pro-life" objection is not an argument against sex selection as such. It is an argument against all forms of embryo screening, including PGD carried out to screen for genetic diseases. Because the pro-life objection finds an overriding moral wrong in the means (namely, the discarding of unwanted embryos), it leaves open the question of whether there is anything wrong with sex selection itself.

The latest sex selection technology poses this question on its own, unclouded by the matter of an embryo's moral status. The Genetics & IVF Institute, a for-profit infertility clinic in Fairfax, Virginia, now offers a sperm-sorting technique that makes it possible for clients to choose the sex of their child before it is conceived. The X-bearing sperm (which produce girls) carry more DNA than Y-bearing sperm (which produce boys); a device called a flow cytometer can separate them. The trademarked process, called MicroSort, has a high rate of success—91 percent for producing girls, 76 percent for boys. The Genetics & IVF Institute licensed the technology from the U.S. Department of Agriculture, which had developed the process for breeding cattle.[18]

If sex selection by sperm sorting is objectionable, it must be for reasons that go beyond the debate about the moral status of the embryo. One such reason is that sex selection is an instrument of sex discrimination, typically against girls, as illustrated by the chilling sex ratios in India and China. And some speculate that societies with substantially more men than women will be less stable, more violent, more prone to crime or war than societies with normal distributions.[19] These are legitimate worries, but the sperm-sorting company has a clever way of addressing them. It offers MicroSort only to couples who want to choose the sex of their child for purposes of family balancing. Those with more sons than daughters can choose a girl, and vice versa. But customers may not use the technology to stock up on children of the same sex, or even to choose the sex of their first-born child. So far, the majority of MicroSort clients have chosen girls.[20]

The case of MicroSort helps us isolate the moral question posed by technologies of enhancement. Put aside familiar debates about safety, embryo loss, and sex discrimination. Imagine that sperm-sorting technologies were employed in a society that did not favor boys over girls, and that wound up with a balanced sex ratio. Would sex selection under those conditions be unobjectionable? What if it became possible to select not only for sex but also for height, eye color, and skin color? What about sexual orientation, IQ, musical ability, and athletic prowess? Or suppose that muscle-enhancement, memory-enhancement, and height-enhancement technologies were perfected to the point where they were safe and available to all. Would they cease to be objectionable?

Not necessarily. In each of these cases, something morally troubling persists. The trouble resides not only in the means but also in the ends being aimed at. It is commonly said that enhancement, cloning, and genetic engineering pose a threat to human dignity. This is true enough. But the challenge is to say *how* these practices diminish our humanity. What aspects of human freedom or human flourishing do they threaten?

2

Bionic Athletes

ONE ASPECT OF our humanity that might be threatened by enhancement and genetic engineering is our capacity to act freely, for ourselves, by our own efforts, and to consider ourselves responsible — worthy of praise or blame — for the things we do and for the way we are. It is one thing to hit seventy home runs as a result of disciplined training and effort, and something else, something less, to hit them with the help of steroids or genetically enhanced muscles. Of course the roles of effort and enhancement will be a matter of degree. But as the role of the enhancement increases, our admiration for the achievement fades. Or rather, our admiration for the achievement shifts from the player to his pharmacist.

THE ATHLETIC IDEAL:
EFFORT VERSUS GIFT

This suggests that our moral response to enhancement is a response to the diminished agency of the person whose achievement is enhanced. The more the athlete relies on drugs or genetic fixes, the less his performance represents his achievement. At the extreme, we might imagine a robotic, bionic athlete who, thanks to implanted computer chips that perfect the angle and timing of his swing, hits every pitch in the strike zone for a home run. The bionic athlete would not be an agent at all; "his" achievements would be those of his inventor. According to this view, enhancement threatens our humanity by eroding human agency. Its ultimate expression is a wholly mechanistic understanding of human action at odds with human freedom and moral responsibility.

Though there is much to be said for this account, I do not think that the main problem with enhancement and genetic engineering is that they undermine effort and erode human agency.[1] The deeper danger is that they represent a kind of hyperagency, a Promethean aspiration to remake nature, including human nature, to serve our purposes and satisfy our desires. The problem is not the drift to mechanism but the drive to mastery. And what the drive to mastery misses, and may even destroy, is an appreciation of the gifted character of human powers and achievements.

To acknowledge the giftedness of life is to recognize that our talents and powers are not wholly our own doing, nor even fully ours, despite the efforts we expend to develop and to exercise them. It is also to recognize that not everything in the world is open to any use we may desire or devise. An appreciation of the giftedness of life constrains the Promethean project and conduces to a certain humility. It is, in part, a religious sensibility. But its resonance reaches beyond religion.

It is difficult to account for what we admire about human activity and achievement without drawing on some version of this idea. Consider two types of athletic achievement: We admire baseball players like Pete Rose, who are not blessed with great natural gifts but who manage, through effort and striving, grit and determination, to excel in their sport. But we also admire players like Joe DiMaggio, whose excellence consists in the grace and effortlessness with which they display their natural gifts. Now suppose we learn that both of those players took performance-enhancing drugs. Whose use of drugs do we find more deeply disillusioning? Which aspect of the athletic ideal—effort or gift—is more deeply offended?

Some might say effort; the problem with drugs is that they provide a shortcut, a way to win without striving. But striving is not the point of sports; excellence is. And excellence consists at least partly in the display of natural talents and gifts that are no doing of the athlete who possesses them. This is an uncomfortable fact for democratic societies. We want to believe that success, in sports and in life, is something we earn, not something we inherit. Natural gifts, and the admiration they inspire, embarrass the meritocratic faith; they cast doubt on the conviction that praise and rewards flow from effort alone. In the face of this embarrassment, we inflate the moral significance of effort and striving, and depreciate giftedness. This distortion can be seen, for example, in television coverage of the Olympics, which focuses less on the feats the athletes perform than on heart-rending stories of the hardships they have overcome, the obstacles they have surmounted, and the struggles they have waged to triumph over an injury, or a difficult upbringing, or political turmoil in their native land.

If effort were the highest athletic ideal, then the sin of enhancement would be the evasion of training and hard work. But effort isn't everything. No one believes that a mediocre basketball player who works and trains even harder than Michael Jordan deserves greater acclaim or a bigger contract. The real problem with genetically altered athletes is that they corrupt athletic competition as a human activity that honors the cultivation and display of natural talents. From this standpoint, enhancement can be seen as the ultimate expression of the ethic of effort and willfulness, a kind of high-tech striving. The ethic of willfulness and the biotechnological powers it now enlists are both arrayed against the claims of giftedness.

Performance Enhancement: High Tech and Low Tech

The line between cultivating natural gifts and corrupting them with artifice may not always be clear. In the beginning, runners ran barefoot. The person who donned the first pair of running shoes may have been accused of tainting the race. The accusation would have been unjust; provided everyone has access to them, running shoes highlight rather than obscure the excellence the race is meant to display. The same cannot be said of all devices athletes employ to improve their performance. When it was discovered that Rosie Ruiz won the 1980 Boston Marathon by slipping away from the pack and riding the subway for part of the race, her prize was withdrawn. The hard cases lie somewhere between running shoes and the subway.

Innovations in equipment are a kind of enhancement, open always to the question of whether they perfect or obscure the skills essential to the game. But bodily enhancement seems to raise the hardest questions. Defenders of enhancement argue that drugs and genetic interventions are no different from other ways athletes alter their bodies to improve their performance, such as with special diets, vitamins, energy bars, over-the-counter supplements, rigorous training regimes, even surgery. Tiger Woods had eyesight so poor he couldn't read the large E on the eye chart. In 1999 he underwent Lasik eye surgery to improve his vision, and he won his next five tournaments.[2]

The remedial nature of the eye surgery makes it

easy to accept. But what if Woods had normal vision and wanted to improve it? Or suppose, as seems to be the case, that the laser treatment gave him better eyesight than the average golfer. Would that make the surgery an illegitimate enhancement?

The answer depends on whether improving the eyesight of golfers is more likely to perfect or to distort the talents and skills that golf at its best is meant to test. The defenders of enhancement are right to this extent: The legitimacy of vision enhancement for golfers does not depend on the means they employ—whether surgery, contact lenses, eye exercises, or copious amounts of carrot juice. If enhancement is troubling because it distorts and overrides natural gifts, the problem is not unique to drugs and genetic alterations; similar objections can also be raised against types of enhancement we commonly accept, such as training and diet.

When, in 1954, Roger Bannister became the first person to break the four-minute mile, his training consisted of a run with friends during his lunch break at the hospital where he worked as a medical student.[3] By the standards of today's training routines, Bannister might as well have been running barefoot. Hoping to improve the performance of American marathon runners, the Nike corporation now sponsors a high-tech training experiment in a hermetically sealed "altitude house" in Portland, Oregon. Molecular filters remove enough oxygen from the house to simulate the thin air found at altitudes of 12,000 to 17,000 feet. Five promising runners have been recruited to live in the house for four to five years, to test the "live high, train low" theory of endurance training. By sleeping at the altitude of the Himalayas, the runners boost their production of oxygen-carrying red blood cells, a key factor in endurance. By working out at sea level—they run more than 100 miles a week—they are able to push their muscles to the maximum. The house is also equipped with devices that monitor the athletes' heart rates, red blood cell counts, oxygen consumption, hormone levels, and brain waves, allowing them to set the time and intensity of their training according to physiological indicators.[4]

The International Olympic Committee is trying to decide whether to ban artificial altitude training. It already prohibits other means by which athletes boost stamina by increasing their concentration of red blood cells, including blood transfusions and injections of erythropoietin (EPO), a hormone produced by the kidneys that stimulates red blood cell production. A synthetic version of EPO, developed to help dialysis patients, has become a popular if illicit performance enhancer for distance runners, cyclists, and cross-country skiers. The IOC instituted testing for EPO use at the Sydney games in 2000, but a new form of EPO gene therapy may prove more difficult to detect than the synthetic version. Scientists working with baboons have found a way to insert a new copy of the gene that produces EPO. Before long, genetically modified runners and cyclists may be able to generate higher-than-normal levels of their own natural EPO for an entire season or longer.[5]

Here is the ethical conundrum: If EPO injections and genetic modifications are objectionable, why isn't Nike's "altitude house" also objectionable? The effect on performance is the same—increasing aerobic endurance by boosting the blood's capacity to carry oxygen to the muscles. It hardly seems nobler to thicken the blood by sleeping in a sealed room with thin air than by injecting hormones or altering one's genes. In 2006 the ethics panel of the World Anti-Doping Agency followed this logic in concluding that the use of low-oxygen chambers and tents (artificial "hypoxic devices") violates "the spirit of sport." This determination brought protests from cyclists, runners, and companies that sell the devices.[6]

If some forms of training are questionable routes to enhanced performance, so are some dietary practices. Over the past thirty years, the size of football players in the NFL has dramatically increased. The average weight of an offensive lineman in the 1972 Super Bowl was an already ample 248 pounds. By 2002 the average Super Bowl lineman weighed 304 pounds, and the Dallas Cowboys boasted the NFL's first 400-pound player, tackle Aaron Gibson, officially listed at 422 pounds. Steroid use no doubt accounted for some of the weight gain among players, especially in the 1970s and 1980s. But steroids were banned in 1990 and the weight increase continued, largely through food intake of gargantuan proportions by linemen eager to make the roster. As Selena Roberts wrote in the *New York Times*, "For some players under intense pressure to add pounds, the science of size comes down to a cocktail of unregulated supplements and a bag of cheeseburgers."[7]

There is nothing high-tech about a mountain of Big Macs. And yet encouraging athletes to use megacalorie diets to turn themselves into 400-pound human shields and battering rams is as ethically questionable as encouraging them to bulk up through the use of steroids, human growth hormone, or genetic alterations. Whatever the means, the push for supersized players is degrading to the game and to the dignity of those who transform their bodies to meet its demands. A retired NFL Hall of Fame lineman laments that the overgrown linemen of today, too big to run sweeps and screens, are capable only of high impact "belly-bumping": "That's all they're doing out there. They are not as athletic, not as quick. They don't use their feet."[8] Enhancing performance by mainlining cheeseburgers does not cultivate athletic excellence but overrides it in favor of a bone-crushing spectacle.

The most familiar argument for banning drugs like steroids is that they endanger athletes' health. But safety is not the only reason to restrict performance-enhancing drugs and technologies. Even enhancements that are safe and accessible to all can threaten the integrity of the game. It is true that, if the rules permitted all manner of drugs, supplements, equipment, and training methods, their use would not constitute cheating. But cheating is not the only way a sport can be corrupted.

Honoring the integrity of a sport means more than playing by the rules, or enforcing them. It means writing the rules in a way that honors the excellences central to the game and rewards the skills of those who play it best.

THE ESSENCE OF THE GAME

Some ways of playing the game, and preparing for it, run the risk of transforming it into something else—something less like a sport and more like a spectacle. A game in which genetically altered sluggers routinely hit home runs might be amusing for a time, but it would lack the human drama and complexity of baseball, in which even the greatest hitters fail more often than they succeed. (Even the fun of watching the annual home-run hitting contest staged by Major League Baseball, a fairly innocent spectacle, presupposes some acquaintance with the real thing—a game in which home runs, far from routine, figure as heroic moments in a larger drama.)

The difference between a sport and a spectacle is the difference between real basketball and "trampoline basketball," in which players can launch themselves high above the basket and dunk the ball; it is the difference between real wrestling and the version staged by the World Wrestling Federation (WWF), in which wrestlers attack their opponents with folding chairs. Spectacles, by isolating and exaggerating through artifice an attention-grabbing feature of a sport, depreciate the natural talents and gifts that the greatest players display. In a game that allowed basketball players to use a trampoline, Michael Jordan's athleticism would no longer loom as large.

Of course, not all innovations in training and equipment corrupt the game. Some, like baseball gloves and graphite tennis rackets, improve it. How can we distinguish changes that improve from those that corrupt? No simple principle can resolve the question once and for all. The answer depends on the nature of the sport, and on whether the new technology highlights or obscures the talents and skills that distinguish the best players. Running shoes improved foot races by reducing the risk that runners would be hampered by contingencies unrelated to the race (like stepping barefoot on a sharp pebble); shoes made the race a truer test of the best runner. Allowing marathon runners to ride the subway to the finish line, or wrestlers to fight with folding chairs, makes a mockery of the skills that marathons and wrestling matches are meant to test.

Arguments about the ethics of enhancement are always, at least in part, arguments about the telos, or point, of the sport in question, and the virtues relevant to the game. This is as true in controversial cases as in obvious ones. Consider coaching. In *Chariots of Fire*, a movie set in 1920s England, authorities at the University of Cambridge chastised one of their star athletes for employing a running coach.[9] Doing so, they argued, violated the spirit of amateur athletics, which included, they thought, training wholly on one's own, or with one's peers. The runner believed that the point of college sports was to develop one's athletic talents as fully as possible, and that the coach would help, not taint, the pursuit of this good. Whether the coach was a legitimate means of performance enhancement depends on which view of the purpose of college sports and their attendant virtues was correct.

Debates about performance enhancement arise in music as well as sports, and take a similar form.

Some classical musicians who suffer from stage fright take beta-blockers to calm their nerves before performing. The drugs, designed to treat cardiac disorders, help nervous musicians by reducing the effect of adrenaline, lowering the heart rate, and enabling them to play unimpeded by quivering hands.[10] Opponents of this practice consider drug-becalmed performance a kind of cheating and argue that part of being a musician is learning to conquer fear the natural way. Defenders of beta-blockers argue that the drugs do not make anyone a better violinist or pianist but simply remove an impediment so that performers can display their true musical gifts. Underlying the debate is a disagreement over the qualities that constitute musical excellence: Is equanimity before a packed house a virtue intrinsic to a great musical performance, or is it merely incidental?

Sometimes mechanical enhancements can be more corrupting than pharmacological ones. Recently, concert halls and opera houses have begun to install sound amplification systems.[11] Music lovers complain that putting mics on the musicians will sully the sound and degrade the art. Great opera singing is not only about hitting the right notes, they argue, but also about projecting the natural human voice to the back of the hall. For classically trained vocalists, projecting one's voice is not simply a matter of cranking up the volume; it is part of the art. The operatic star Marilyn Horne calls sound enhancement the "kiss of death for good singing."[12]

Anthony Tommasini, classical music critic for the *New York Times*, describes how sound amplification transformed, and in some ways degraded, the Broadway musical: "In its thrilling early decades the Broadway musical was a bracingly literate genre in which clever words were mixed in ingenious ways with snappy, snazzy, or wistfully tuneful music. In its essence, though, it was a word-driven art form. . . . But when amplification took hold on Broadway, audiences inevitably grew less alert, more passive. It began changing every element of the musical, from the lyrics (which grew less subtle and intricate), to the subject matter and musical styles (the bigger, the plusher, the schlockier, the better)." As musicals became "less literate and more obvious," singers with voices of "operatic dimensions became marginalized," and the genre devolved into melodramatic spectacles like *Phantom of the Opera* and *Miss Saigon.* As the musical has adapted to amplification, "the art form has diminished, or at least become something different."[13]

Fearing that opera may suffer a similar fate, Tommasini wishes that traditional, unamplified opera could be preserved, as an option, alongside the electronically enhanced version. This suggestion recalls proposals for parallel sports competitions for the enhanced and the unenhanced. One such proposal was offered by an enhancement enthusiast writing in *Wired,* a technology magazine: "Create one league for the genetically engineered home-run hitter and another for the human-scale slugger. One event for the sprinter pumped up on growth hormones and another for the free-range slowpoke." The writer was convinced that the bulked-up leagues would draw higher television ratings than their all-natural counterparts.[14]

Whether amplified and traditional opera, or bulked-up and "free-range" sports leagues, could coexist for long is difficult to say. In art as in sports, technologically enhanced versions of a practice seldom leave old ways undisturbed; norms change, audiences become rehabituated, and spectacle exerts a certain allure, even as it deprives us of unadulterated access to human talents and gifts.

Assessing the rules of athletic competition for their fit with the excellences essential to the sport will strike some as unduly judgmental, reminiscent of the arch, aristocratic sensibility of the Cambridge dons in *Chariots of Fire*. But it is difficult to make sense of what we admire about sports without making some judgment about the point of the game and its relevant virtues.

Consider the alternative. Some people deny that sports have a point. They reject the idea that the rules of a game should fit the telos of the sport, and honor the talents displayed by those who play it well. According to this view, the rules of any game are wholly arbitrary, justified only by the entertainment they provide and the number of spectators they attract. The clearest statement of this view appears, of all places, in a U.S. Supreme Court opinion by Justice Antonin Scalia. The case involved a professional golfer who, unable to walk without pain due to a congenital leg disease, sued under the Americans with Disabilities Act for the right to use a golf cart in professional tournaments. The Supreme Court held in his favor, reasoning that walking the course was not an essential aspect of golf. Scalia dissented, arguing that it is impossible to distinguish essential from incidental features of a game: "To say that something is 'essential' is ordinarily to say that it is necessary to the achievement of a certain object. But since it is the very nature of a game to have no object except amusement (that is what distinguishes games from productive activity), it is quite impossible to say that any of a game's arbitrary rules is 'essential.'" Since the rules of golf "are (as in all games) entirely arbitrary," Scalia argued, there is no basis for critically assessing the rules laid down by the association that governs the game.[15]

But Scalia's view of sports is far-fetched. It would strike any sports fan as odd. If people really believed that the rules of their favorite sport were arbitrary rather than designed to call forth and celebrate certain talents and virtues worth admiring, they would find it difficult to care about the outcome of the game.[16] Sport would fade into spectacle, a source of amusement rather than a subject of appreciation. Safety considerations aside, there would be no reason to restrict performance-enhancing drugs and genetic alterations—no reason, at least, tied to the integrity of the game rather than the size of the crowd.

The descent of sport into spectacle is not unique to the age of genetic engineering. But it illustrates how performance-enhancing technologies, genetic or otherwise, can erode the part of athletic and artistic performance that celebrates natural talents and gifts.

3

Designer Children, Designing Parents

THE ETHIC OF GIFTEDNESS, under siege in
sports, persists in the practice of parenting. But
here, too, bioengineering and genetic enhancement
threaten to dislodge it. To appreciate children as
gifts is to accept them as they come, not as objects
of our design, or products of our will, or instru
ments of our ambition. Parental love is not contingent on the talents and attributes the child hap-
pens to have. We choose our friends and spouses at
least partly on the basis of qualities we find attrac-
tive. But we do not choose our children. Their
qualities are unpredictable, and even the most con-
scientious parents cannot be held wholly responsible for the kind of child they have. That is why
parenthood, more than other human relationships,
teaches what the theologian William F. May calls
an "openness to the unbidden."[1]

MOLDING AND BEHOLDING

May's resonant phrase describes a quality of character and heart that restrains the impulse to mastery and control and prompts a sense of life as gift. It helps us see that the deepest moral objection to enhancement lies less in the perfection it seeks than in the human disposition it expresses and promotes. The problem is not that the parents usurp the autonomy of the child they design. (It is not as if the child could otherwise choose her genetic traits for herself.) The problem lies in the hubris of the designing parents, in their drive to master the mystery of birth. Even if this disposition does not make parents tyrants to their children, it disfigures the relation between parent and child, and deprives the parent of the humility and enlarged human sympathies that an openness to the unbidden can cultivate.

To appreciate children as gifts or blessings is not to be passive in the face of illness or disease. Healing a sick or injured child does not override her natural capacities but permits them to flourish. Although medical treatment intervenes in nature, it does so for the sake of health, and so does not represent a boundless bid for mastery and dominion. Even strenuous attempts to treat or cure disease do not constitute a Promethean assault on the given. The reason is that medicine is governed, or at least guided, by the norm of restoring and preserving the natural human functions that constitute health.

Medicine, like sports, is a practice with a purpose, a telos, that orients and constraints it. Of course what counts as good health or normal human functioning is open to argument; it is not only a biological question. People disagree, for example, about whether deafness is a disability to be cured or a form of community and identity to be cherished. But even the disagreement proceeds from the assumption that the point of medicine is to promote health and cure disease.

Some people argue that a parent's obligation to heal a sick child implies an obligation to enhance a healthy one, to maximize his or her potential for success in life. But this is true only if one accepts the utilitarian idea that health is not a distinctive human good, but simply a means of maximizing happiness or well-being. Bioethicist Julian Savulescu argues, for example, that "health is not intrinsically valuable," only "instrumentally valuable," a "resource" that allows us to do what we want. This way of thinking about health rejects the distinction between healing and enhancing. According to Savulescu, parents not only have a duty to promote their children's health; they are also "morally obliged to genetically modify their children." Parents should use technology to manipulate their children's "memory, temperament, patience, empathy, sense of humor, optimism," and other characteristics in order to give them "the best opportunity of the best life."[2]

But it is a mistake to think of health in wholly instrumental terms, as a way of maximizing something else. Good health, like good character, is a constitutive element of human flourishing. Although more health is better than less, at least within a certain range, it is not the kind of good that can be maximized. No one aspires to be a virtuoso at health (except, perhaps, a hypochondriac). During the 1920s, eugenicists held health contests at state fairs and awarded prizes to the "fittest families." But this bizarre practice illustrates the folly of conceiving health in instrumental terms, or as a good to be maximized. Unlike the talents and traits that bring success in a competitive society, health is a bounded good; parents can seek it for their chil-

dren without risk of being drawn into an ever-esca-
lating arms race.

In caring for the health of their children, parents
do not cast themselves as designers or convert their
children into products of their will or instruments
of their ambition. The same cannot be said of parents who pay large sums to select the sex of their
child (for nonmedical reasons) or who aspire to
bioengineer their child's intellectual endowments
or athletic prowess. Like all distinctions, the line
between therapy and enhancement blurs at the
edges. (What about orthodontics, for example, or
growth hormone for very short kids?) But this does
not obscure the reason the distinction matters: parents bent on enhancing their children are more
likely to overreach, to express and entrench attitudes at odds with the norm of unconditional love.

Of course, unconditional love does not require
that parents refrain from shaping and directing the
development of their child. To the contrary, parents have an obligation to cultivate their children,
to help them discover and develop their talents and
gifts. As May points out, parental love has two aspects: accepting love and transforming love. Accepting love affirms the being of the child, whereas

transforming love seeks the well-being of the child. Each side of parental love corrects the excesses of the other: "Attachment becomes too quietistic if it slackens into mere acceptance of the child as he is." Parents have a duty to promote their child's excellence.[3]

These days, however, overly ambitious parents are prone to get carried away with transforming love—promoting and demanding all manner of accomplishments from their children, seeking perfection. "Parents find it difficult to maintain an equilibrium between the two sides of love," May observes. "Accepting love, without transforming love, slides into indulgence and finally neglect. Transforming love, without accepting love, badgers and finally rejects." May finds in these competing impulses a parallel with modern science; it, too, engages us in beholding the given world, studying and savoring it, and also in molding the world, transforming and perfecting it.[4]

The mandate to mold our children, to cultivate and improve them, complicates the case against enhancement. We admire parents who seek the best for their children, who spare no effort to help them achieve happiness and success. What, then, is the difference between providing such help through education and training and providing it by means of genetic enhancement? Some parents confer advantages on their children by enrolling them in expensive schools, hiring private tutors, sending them to tennis camp, providing them with piano lessons, ballet lessons, swimming lessons, SAT prep courses, and so on. If it is permissible, even admirable, for parents to help their children in these ways, why isn't it equally admirable for parents to use whatever genetic technologies may emerge (provided they are safe) to enhance their child's intelligence, musical ability, or athletic skill?

Defenders of enhancement argue that there is no difference, in principle, between improving children through education and improving them through bioengineering. Critics of enhancement insist there is all the difference in the world. They argue that trying to improve children by manipulating their genetic makeup is reminiscent of eugenics, the discredited movement of the past century to improve the human race through policies (including forced sterilization and other odious measures) aimed at improving the gene pool. These competing analogies help clarify the moral status of genetic enhancement. Is the attempt of parents to enhance their children through genetic engineering more like education and training (a presumably good thing) or more like eugenics (a presumably bad thing)?

The defenders of enhancement are right to this extent: Improving children through genetic engineering is similar in spirit to the heavily managed, high-pressure child-rearing practices that have become common these days. But this similarity does not vindicate genetic enhancement. On the contrary, it highlights a problem with the trend toward hyperparenting.[5] The most conspicuous examples are sports-crazed parents bent on making champions of their children. Sometimes they are successful, as in the case of Richard Williams, who reportedly planned the tennis careers of his daughters, Venus and Serena Williams, before they were born; or Earl Woods, who handed a golf club to young Tiger Woods while he was still in a playpen. "Let's face it, no kid puts themselves into a sport this way," Richard Williams told the *New York Times.* "The parents do it, and I'm guilty there. If you don't plan it, believe me, it's not going to happen."[6]

A similar sentiment can be found outside the ranks of elite sports, among the overwrought parents on the sidelines of soccer fields and Little League diamonds across the land. So acute is the epidemic of parental intrusiveness and competitiveness that youth sports leagues have sought to control it by establishing parent-free zones, silent weekends (no yelling or cheering), and awards for parental sportsmanship and restraint.[7]

Hectoring from the sidelines is not the only toll that hyperparenting takes on young athletes. As pickup games and playground sports have given way to sports leagues organized and managed by driving parents, pediatricians report an alarming increase in overuse injuries among teenagers. Today, sixteen-year-old pitchers are undergoing elbow reconstruction surgery, a procedure once performed only on major league pitchers seeking to prolong their careers. Dr. Lyle Micheli, the director of sports medicine at Boston Children's Hospital, reports that 70 percent of the young patients he treats suffer from overuse injuries, up from 10 percent twenty-five years ago. Sports doctors attribute the epidemic of overuse injuries to the growing tendency to have children specialize in a single sport from an early age, and to train for it year-round. "Parents think they are maximizing their child's chances by concentrating on one sport," said Dr. Micheli. "The results are often not what they expected."[8]

Youth sports officials and doctors are not the only ones seeking ways to rein in overbearing parents. College administrators also complain of a growing problem with parents eager to control their children's lives—writing their children's college applications, phoning to badger the admissions office, helping write term papers, staying overnight in dorm rooms. Some parents even call college officials to ask that their child be awakened in the morning.[9] "Parents of college students are out of control," says Marilee Jones, dean of admissions at MIT, who has made a mission of urging anxious parents to back off.[10] Judith R. Shapiro, president of Barnard College, agrees. In an op-ed titled "Keeping Parents off Campus," she wrote: "Their sense of entitlement as consumers, along with an inability to let go, leads some parents to want to manage all aspects of their children's college lives—from the quest for admission to their choice of major. Such parents, while the exception, are nonetheless an increasing fact of life for faculty, deans and presidents."[11]

The frenzied drive by parents to mold and manage their children's academic careers has intensified over the past decade as baby boomers, accustomed to having control, prepare to send their kids to college. A generation ago, few high school students bothered to prepare for the SAT. Today parents spend large sums on commercial SAT prep courses, tutors, books, and software for their college-bound children, making test preparation a $2.5 billion industry.[12] From 1992 to 2001 Kaplan, a leading test prep company, saw its gross revenues increase 225 percent.[13]

SAT prep courses are not the only way the anxious affluent try to polish and package their college-bound progeny. Educational psychologists report that growing numbers of parents seek to have their high school junior or senior diagnosed with a learning disability, solely for the purpose of receiving additional time while taking the SAT. This "diagnosis-shopping" was apparently spurred by the College Board's announcement in 2002 that it would no longer place an asterisk next to the scores of students allowed extra time owing to a learning disability. Parents shell out as much as $2,400 for an evaluation and $250 an hour for a psychologist to testify on the student's behalf at the high school or the Educational Testing Service, which produces the SAT. If one psychologist does not produce the diagnosis they want, they take their business else-where.[14]

Hyperparenting is strenuous and time consum-ing, so some parents subcontract the job to private counselors and consultants. For fees of up to $500 an hour, private college admissions counselors guide students through the rigors of the application process—deciding where to apply, editing admis-sions essays, compiling résumés, practicing for in-terviews. Mounting parental angst has made the private counseling business a growth industry. Ac-cording to the Independent Educational Consul-tants Association, which represents the profession, more than 10 percent of today's college freshman have used paid counselors, up from 1 percent in 1990.[15]

The most upscale firm in the business, IvyWise in Manhattan, offers a two-year "platinum pack-age" of college admissions help for $32,995.[16] For this handsome fee, Katherine Cohen, the founder of the firm, starts early with her clients and tells them what extracurricular activities, volunteer work, and summer experiences they should undertake in high school to burnish their résumé and boost their chances of admission. She not only markets kids to colleges but also helps with the product development—a hyperparent for hire. "I don't guide applications," Cohen says. "I guide lives."[17]

For some parents, the scramble to package and position their children for admission to an elite college begins in early childhood. Cohen's partner offers a service called IvyWise Kids that caters to parents eager to win spots for their children in the most coveted private elementary schools in New York City (the so-called Baby Ivies), and in the highly sought nursery schools that feed into them.[18] The crazed competition for preschool admissions was highlighted a few years ago by the story of Jack Grubman, a Wall Street stock analyst. He claimed in an e-mail to have upgraded his rating of AT&T stock in order to curry favor with his boss, who was helping to get Grubman's twin two-year-old daughters admitted to the prestigious 92nd Street Y nursery school.[19]

THE PRESSURE TO PERFORM

Grubman's willingness to move heaven and earth, and even the market, to get his two-year-olds into a fancy nursery school is a sign of the times. It tells of mounting pressures in American life that are changing the expectations parents have for their children and increasing the demands placed on children to perform. When preschoolers apply to private kindergartens and elementary schools, their fate depends on favorable letters of recommendation and a standardized test intended to measure their intelligence and development. Some parents have their four-year-olds coached to prepare for the test. Many also shell out $34.95 for a new big-selling toy called the Time Tracker, a brightly colored device with lights and a digital panel designed to teach young children how to keep time during standardized tests. Recommended for children age four and above, the Time Tracker features a helpful electronic male voice that announces "Begin" and "Time's up."[20]

The testing of toddlers is not restricted to private schools. The Bush administration mandated that all four-year-olds enrolled in Head Start take standardized tests. Increased state testing in elementary grades has led school districts around the country to tighten curricula in kindergarten, where reading, math, and science are displacing art, recess, and naptime. By the time kids reach first and second grade, they must now contend with homework and heavy backpacks. Between 1981 and 1997, the amount of homework assigned to children six to eight years old tripled.[21]

As the pressure for performance increases, so does the need to help distractible children concentrate on the task at hand. Some attribute the sharp increase in diagnoses for attention deficit and hyperactivity disorder (ADHD) to the new demands placed on children to perform. Dr. Lawrence Diller, a pediatrician and the author of *Running on Ritalin*, estimates that 5 to 6 percent of American children under eighteen (a total of four to five million kids) are currently prescribed Ritalin and other stimulants, the treatment of choice for ADHD. (Stimulants counteract hyperactivity by making it easier for children to focus and sustain their attention and stop flitting from one thing to another.) Over the past fifteen years, the legal production of Ritalin increased by 1,700 percent, and the production of the amphetamine Adderall, also marketed for treatment of ADHD, rose 3,000 percent. For the pharmaceutical companies, the U.S. market for Ritalin and related drugs is a bonanza: one billion dollars a year.[22]

While Ritalin prescriptions for children and adolescents have skyrocketed in recent years, not all users suffer from attention disorders or hyperactivity. High school and college students have learned that prescription stimulants improve concentration in those with a normal attention span; some buy or borrow their classmates' Ritalin to enhance their performance on the SAT or on college exams. One of the most troubling findings about Ritalin use is that doctors increasingly prescribe it for preschoolers. Although the drug is not approved for children under six years old, prescription rates for two- to four-year-old children nearly tripled from 1991 to 1995.[23]

Since Ritalin works for both medical and nonmedical purposes—to remedy ADHD and to enhance the performance of healthy kids seeking a competitive edge—it raises the same moral dilemmas posed by other technologies of enhancement. However those dilemmas are resolved, the debate over Ritalin reveals the cultural distance we have traveled since the debate over drugs (such as marijuana and LSD) a generation ago. Unlike the drugs of the sixties and seventies, Ritalin and Adderall are not for checking out but for buckling down, not for beholding the world and taking it in, but for molding the world and fitting in. We used to speak of nonmedical drug use as "recreational." That term no longer applies. The steroids and stimulants that figure in the enhancement debate are not a source of recreation but a bid for compliance, a way of answering a competitive society's demand to improve our performance and perfect our nature. This demand for performance and perfection animates the impulse to rail against the given. It is the deepest source of the moral trouble with enhancement.

Some see a bright line between genetic enhancement and other ways that people seek improvement in their children and themselves. Genetic manipulation seems somehow worse—more intrusive, more sinister—than other ways of enhancing performance and seeking success. But morally speaking, the difference is less significant than it seems.

Those who argue that bioengineering is similar in spirit to other ways ambitious parents shape and mold their children have a point. But this similarity does not give us reason to embrace the genetic manipulation of children. Instead, it gives us reason to question the low-tech, high-pressure child-rearing practices we commonly accept. The hyperparenting familiar in our time represents an anxious excess of mastery and dominion that misses the sense of life as gift. This draws it disturbingly close to eugenics.

4

The Old Eugenics and the New

EUGENICS WAS A MOVEMENT of large ambition—to improve the genetic makeup of the human race. The term, which means "well born," was coined in 1883 by Sir Francis Galton, a cousin of Charles Darwin, who applied statistical methods to the study of heredity.[1] Persuaded that heredity governed talent and character, he thought it possible "to produce a highly gifted race of men by judicious marriages during several consecutive generations."[2] He called for eugenics to be "introduced into the national conscience, like a new religion," encouraging the talented to choose their mates with eugenic aims in mind. "What nature does blindly, slowly, and ruthlessly, man may do providently, quickly, and kindly. . . . The improvement of our stock seems to me one of the highest objects that we can reasonably attempt."[3]

THE OLD EUGENICS

Galton's idea spread to America, where it fueled a popular movement in the early decades of the twentieth century. In 1910, biologist and eugenic crusader Charles B. Davenport opened the Eugenic Records Office in Cold Spring Harbor, Long Island. Its mission was to send fieldworkers into prisons, hospitals, almshouses, and insane asylums across the country to investigate and collect data on the genetic backgrounds of so-called defectives. In Davenport's words, the project was to catalog "the great strains of human protoplasm that are coursing through the country."[4] Davenport hoped such data would provide the basis for eugenic efforts to prevent reproduction of the genetically unfit.

The crusade to rid the nation of defective protoplasm was no marginal movement of racists and cranks. Davenport's work was funded by the Carnegie Institution; Mrs. E. H. Harriman, widow and heir of the Union Pacific railroad magnate; and John D. Rockefeller, Jr. Leading progressive reformers of the day rallied to the eugenic cause. Theodore Roosevelt wrote Davenport: "Some day, we will realize that the prime duty, the inescapable duty, of the good citizen of the right type, is to leave his or her blood behind him in the world; and that we have no business to permit the perpetuation of citizens of the wrong type."[5] Margaret Sanger, pioneering feminist and advocate of birth control, also embraced eugenics: "More children from the fit, less from the unfit—that is the chief issue of birth control."[6]

Part of the eugenic program was hortatory and educational. The American Eugenics Society sponsored "Fitter Families" contests at state fairs around the country, alongside the livestock competitions. Contestants submitted their eugenic histories and underwent medical, psychological, and intelligence testing, and the fittest families were awarded trophies. By the 1920s, eugenics courses were offered at 350 of the nation's colleges and universities, alerting privileged young Americans to their reproductive duty.[7]

But the eugenics movement also had a harsher face. Eugenics advocates lobbied for legislation to prevent those with undesirable genes from reproducing, and in 1907 Indiana adopted the first law providing for the forced sterilization of mental patients, prisoners, and paupers. Twenty-nine states ultimately adopted forced-sterilization laws, and more than 60,000 genetically "deficient" Americans were sterilized.

In 1927 the U.S. Supreme Court upheld the constitutionality of sterilization laws in the notorious case of *Buck v. Bell*. The case involved Carrie Buck, a seventeen-year-old unwed mother who had been committed to a Virginia home for the feebleminded and ordered to undergo sterilization. Justice Oliver Wendell Holmes wrote the opinion for the eight-to-one majority upholding the sterilization law: "We have seen more than once that the public welfare may call upon the best citizens for their lives. It would be strange if it could not call upon those who already sap the strength of the State for these lesser sacrifices. . . . The principle that sustains compulsory vaccination is broad enough to cover cutting the Fallopian tubes. It is better for all the world, if instead of waiting to execute degenerate offspring for crime, or to let them starve for their imbecility, society can prevent those who are manifestly unfit from continuing their kind." Referring to the fact that Carrie Buck's mother and, allegedly, her daughter were also found to be mentally deficient, Holmes concluded: "Three generations of imbeciles are enough."[8]

In Germany, America's eugenic legislation found

an admirer in Adolf Hitler. In *Mein Kampf* he offered a statement of the eugenic faith: "The demand that defective people be prevented from propagating equally defective offspring is a demand of the clearest reason and, if systematically executed, represents the most humane act of mankind. It will spare millions of unfortunates undeserved sufferings, and consequently will lead to a rising improvement of health as a whole."[9] When he seized power in 1933, Hitler issued a far-reaching eugenic sterilization law that drew praise from American eugenicists. The *Eugenical News*, a publication of Cold Spring Harbor, published a verbatim translation of the law and proudly noted its similarities to the model sterilization law proposed by the American eugenics movement. In California, where eugenic sentiment ran high, the *Los Angeles Times* magazine published an upbeat account of Nazi eugenics in 1935. "Why Hitler Says: 'Sterilize the Unfit!'" ran the buoyant headline. "Here, perhaps, is an aspect of the new Germany that America, with the rest of the world, can little afford to criticize."[10]

Ultimately, Hitler carried eugenics beyond sterilization to mass murder and genocide. By the end of World War II, news of the Nazi atrocities contributed to the retreat of the American eugenics movement. Involuntary sterilizations declined in the 1940s and '50s, though some states continued to perform them into the 1970s. In 2002 and 2003, after journalistic investigations brought past eugenic cruelties to the public's attention, the governors of Virginia, Oregon, California, North Carolina, and South Carolina issued formal apologies to victims of forced sterilization.[11]

The shadow of eugenics hangs over today's debates about genetic engineering and enhancement. Critics of genetic engineering argue that human cloning, enhancement, and the quest for designer children are nothing more than "privatized" or "free-market" eugenics. Defenders of enhancement reply that genetic choices freely made are not really eugenic, at least not in the pejorative sense that term conveys. To remove the coercion, they argue, is to remove the very thing that makes eugenic policies repugnant.

Sorting out the lesson of eugenics is another way of wrestling with the ethics of enhancement. The Nazis gave eugenics a bad name. But what exactly was wrong with it? Is eugenics objectionable only insofar as it is coercive? Or is there something wrong with even noncoercive ways of controlling the genetic makeup of the next generation?

FREE-MARKET EUGENICS

Consider a recent eugenics policy that stops short of coercion. In the 1980s, Lee Kuan Yew, the prime minister of Singapore, was worried that well-educated Singaporean women were producing fewer children than less-educated ones. "If we continue to reproduce ourselves in this lopsided way," he said, "we will be unable to maintain our present standards." Subsequent generations, he feared, would become "depleted of the talented."[12] To stave off decline, the government instituted policies to encourage college graduates to marry and have children—a state-run computer dating service, financial incentives for educated women to bear children, courtship classes in the undergraduate curriculum, and free "love boat" cruises for single college graduates. At the same time, low-income women who lacked a high school degree were offered $4,000 as a down payment on a low-cost apartment—provided they were willing to be sterilized.[13]

Singapore's policy gave eugenics a free-market twist; rather than force disfavored citizens to undergo sterilization, it paid them to do so. But those who find traditional eugenic schemes morally abhorrent are likely to be troubled by Singapore's voluntary version as well. Some might object that the $4,000 inducement is akin to coercion, especially for poor women with limited life prospects. Others might object that even the love-boat cruises for the privileged are part of a collectivist program that intrudes on reproductive choices that people should be free to make for themselves, without the heavy hand or watchful eye of the state. (The policies were reportedly unpopular among women, who resented being urged to "breed" for Singapore.)[14] But eugenics is also objectionable on other grounds; even where no coercion is involved, there is something wrong with the ambition, be it individual or collective, to determine the genetic characteristics of our progeny by deliberate design. These days, this ambition is less likely to be found in state-sponsored eugenics policies than in procreative practices that enable parents to pick and choose the kind of children they will have.

James Watson, the biologist who, with Francis Crick, discovered the double-helix structure of

DNA, sees nothing wrong with genetic engineering and enhancement, provided they are freely chosen rather than state-imposed. And yet, for Watson, the language of choice coexists with the old eugenic sensibility. "If you really are stupid, I would call that a disease," Watson recently told the *Times* of London. "The lower 10 per cent who really have difficulty, even in elementary school, what's the cause of it? A lot of people would like to say, 'Well, poverty, things like that.' It probably isn't. So I'd like to get rid of that, to help the lower 10 per cent."[15]

A few years earlier, Watson had stirred controversy by saying that, if a gene for homosexuality were discovered, a pregnant woman who did not want a homosexual child should be free to abort a fetus that carried it. When his remark provoked an uproar, he replied that he was not singling out gays but asserting a principle: women should be free to abort fetuses for any reason of genetic preference — whether testing showed the child would be born dyslexic or lacking musical talent or too short to play basketball.[16]

Watson's scenarios pose no special challenge to pro-life opponents of abortion, for whom all abortion is an unspeakable crime. But for those who do not subscribe to the right-to-life position, Watson's scenarios raise a hard question: If it is morally troubling to contemplate abortion to avoid a gay child or a dyslexic one, doesn't this suggest there is something wrong with acting on eugenic preferences, even where no coercion is involved?

Or consider the market in eggs and sperm. Artificial insemination allows prospective parents to shop for gametes with the genetic traits they desire in their offspring. It is a less predictable way to design children than cloning or preimplantation genetic diagnosis. But it offers a good example of a procreative practice in which the old eugenics meets the new consumerism. Recall the ad that appeared in some Ivy League college newspapers, offering $50,000 for an egg from a young woman who was at least five feet, ten inches tall, athletic, without major family medical problems, and with a combined SAT score of 1400 or above. More recently, a Web site was launched claiming to auction eggs from fashion models whose photos appeared on the site—at starting bids of $15,000 to $150,000.[17]

On what grounds, if any, is the egg market morally objectionable? Since no one is forced to buy or sell, it cannot be wrong for reasons of coercion.

Some might worry that hefty prices would exploit poor women by presenting them with an offer they could not afford to refuse. But the designer eggs that fetch the highest prices are likely to be sought from the privileged, not the poor. If the market for premium eggs gives us moral qualms, it shows that eugenic concerns are not put to rest by freedom of choice.

A tale of two sperm banks helps explain why. The Repository for Germinal Choice, one of America's first sperm banks, was not a commercial enterprise. It was opened in 1980 by Robert Graham, a eugenic philanthropist dedicated to improving the world's "germ plasm" and counteracting the rise of "retrograde humans."[18] His plan was to collect the sperm of Nobel Prize–winning scientists and make it available to women seeking donors, in the hope of breeding supersmart babies. But Graham had trouble persuading Nobel Prize winners to donate their sperm to his bizarre scheme, and so settled for sperm from young scientists of high promise. The sperm bank closed in 1999.[19]

By contrast, California Cryobank, one of the world's leading sperm banks, is a for-profit company. It has no eugenic mission.[20] Dr. Cappy Rothman, cofounder of the firm, has nothing but disdain for Graham's eugenics. And yet the standards Cryobank imposes on the sperm donors it recruits are no less exacting than Graham's. Cryobank has offices in Cambridge, Massachusetts, located between Harvard and MIT, and in Palo Alto, California, near Stanford. It advertises for donors in campus newspapers (and offers to pay up to $900 per month), and accepts fewer than 3 percent of the donors who apply.

Cryobank's marketing materials play up the prestigious source of its sperm. Its donor catalog provides detailed information about the physical characteristics of each donor, as well as his ethnic origin and college major. For an extra fee, prospective customers can buy the results of a test that assesses the donor's temperament and character type. Rothman reports that Cryobank's ideal sperm donor has a college degree, is six feet tall, and has brown eyes, blond hair, and dimples—not because the company wants to propagate those traits, but because those are the traits his customers want. "If our customers wanted high-school dropouts, we would give them high-school dropouts."[21]

Not everyone objects to marketing sperm. But anyone who is troubled by the eugenic aspect of the Nobel Prize sperm bank should be equally troubled by Cryobank, consumer-driven though it be. What, after all, is the moral difference between designing children according to an explicit eugenic purpose and designing children according to the dictates of the market? Whether the aim is to improve humanity's "germ plasm" or to cater to consumer preferences, both practices are eugenic insofar as both make children into products of deliberate design.

Liberal Eugenics

In the age of the genome, the language of eugenics is making a comeback, not only among critics but also among defenders of enhancement. An influential school of Anglo-American political philosophers calls for a new "liberal eugenics," by which they mean noncoercive genetic enhancements that do not restrict the autonomy of the child. "While old-fashioned authoritarian eugenicists sought to produce citizens out of a single centrally designed mould," writes Nicholas Agar, "the distinguishing mark of the new liberal eugenics is state neutrality."[22] Governments may not tell parents what sort of children to design, and parents may engineer in their children only those traits that improve their capacities without biasing their choice of life plans.

A recent text on genetics and justice, written by bioethicists Allen Buchanan, Dan W. Brock, Norman Daniels, and Daniel Wikler, offers a similar view: The "bad reputation of eugenics" is due to practices that "might be avoidable in a future eugenic program." The problem with the old eugenics was that its burdens fell disproportionately on the weak and the poor, who were unjustly segregated and sterilized. But provided that the benefits and burdens of genetic improvement are fairly distributed, these bioethicists argue, eugenic measures are unobjectionable and may even be morally required.[23]

The legal philosopher Ronald Dworkin also defends a liberal version of eugenics. There is nothing wrong with the ambition "to make the lives of future generations of human beings longer and more full of talent and hence achievement," Dworkin writes. "On the contrary, if playing God means struggling to improve our species, bringing into our conscious designs a resolution to improve what God deliberately or nature blindly has evolved over eons, then the first principle of ethical individualism commands that struggle."[24] The libertarian philosopher Robert Nozick proposed a "genetic supermarket" that would enable parents to order children by design without imposing a single design on the society as a whole: "This supermarket system has the great virtue that it involves no centralized decision fixing the future human type(s)."[25]

Even John Rawls, in his classic work, *A Theory of Justice* (1971), offered a brief endorsement of liberal eugenics. Even in a society that agrees to share the benefits and burdens of the genetic lottery, Rawls wrote, it is "in the interest of each to have greater natural assets. This enables him to pursue a preferred plan of life." The parties to the social contract "want to insure for their descendants the best genetic endowment (assuming their own to be fixed)." Eugenic policies are therefore not only permissible but required as a matter of justice. "Thus over time a society is to take steps at least to preserve the general level of natural abilities and to prevent the diffusion of serious defects."[26]

While liberal eugenics is a less dangerous doctrine than the old eugenics, it is also less idealistic. For all its folly and darkness, the eugenics movement of the twentieth century was born of the aspiration to improve humankind, or to promote the collective welfare of entire societies. Liberal eugenics shrinks from collective ambitions. It is not a movement of social reform but rather a way for privileged parents to have the kind of children they want and to arm them for success in a competitive society.

But despite its emphasis on individual choice, liberal eugenics implies more state compulsion than first appears.[27] Defenders of enhancement see no moral difference between improving a child's intellectual capacities through education and doing so through genetic alteration. All that matters, from the liberal-eugenics standpoint, is that neither the education nor the genetic alteration violates the child's autonomy, or "right to an open future."[28] Provided the enhanced capacity is an "all-purpose" means, and so does not point the child toward any particular career or life plan, it is morally permissible.

However, given the duty of parents to promote the well-being of their children (while respecting their right to an open future), such enhancement becomes not only permissible but obligatory. Just as the state can require parents to send their children to school, so it can require parents to use genetic technologies (provided they are safe) to boost their child's IQ. What matters is that the capacities being enhanced are "general-purpose means, useful in carrying out virtually any plan of life. . . . The closer such capacities are to truly all-purpose means, the less objection there should be to the state encouraging or even requiring genetic enhancements of those capabilities."[29] Properly understood, the liberal "principle of ethical individualism" not only permits but "commands the struggle" to "make the lives of future generations of human beings longer and more full of talent and hence achievement."[30] So liberal eugenics does not reject state-imposed genetic engineering after all; it simply requires that the engineering respect the autonomy of the child being designed.

Although liberal eugenics finds support among many Anglo-American moral and political philosophers, Jürgen Habermas, Germany's most prominent political philosopher, opposes it. Acutely aware of Germany's dark eugenic past, Habermas argues against the use of embryo screening and genetic manipulation for nonmedical enhancement. His case against liberal eugenics is especially intriguing because he believes it rests wholly on liberal premises and need not invoke spiritual or theological notions. His critique of genetic engineering "does not relinquish the premises of postmetaphysical thinking," by which he means it does not depend on any particular conception of the good life. Habermas agrees with John Rawls that, since people in modern pluralist societies disagree about morality and religion, a just society should not take sides in such disputes but should instead accord each person the freedom to choose and pursue his or her own conception of the good life.[31]

Genetic intervention to select or improve children is objectionable, Habermas argues, because it violates the liberal principles of autonomy and equality. It violates autonomy because genetically programmed persons cannot regard themselves as "the sole authors of their own life history."[32] And it undermines equality by destroying "the essentially symmetrical relations between free and equal human beings" across generations.[33] One measure of this asymmetry is that, once parents become the designers of their children, they inevitably incur a responsibility for their children's lives that cannot possibly be reciprocal.[34]

Habermas is right to oppose eugenic parenting, but wrong to think that the case against it can rest on liberal terms alone. The defenders of liberal eugenics have a point when they argue that designer children are no less autonomous with respect to their genetic traits than children born the natural way. It is not as if, absent eugenic manipulation, we can choose our genetic inheritance for ourselves. As for Habermas's worry about equality and reciprocity between the generations, defenders of liberal eugenics can reply that this worry, though legitimate, does not apply uniquely to genetic manipulation. The parent who forces her child to practice the piano incessantly from the age of three, or to hit tennis balls from dawn to dusk, also exerts a kind of control over the child's life that cannot possibly be reciprocal. The question, liberals insist, is whether the parental intervention, be it eugenic or environmental, undermines the child's freedom to choose her own life plan.

An ethic of autonomy and equality cannot explain what is wrong with eugenics. But Habermas has a further argument that cuts deeper, even as it points beyond the limits of liberal, or "postmetaphysical" considerations. This is the idea that "we experience our own freedom with reference to something which, by its very nature, is not at our disposal." To think of ourselves as free, we must be able to ascribe our origins "to a beginning which eludes human disposal," a beginning that arises from "something—like God or nature—that is not at the disposal of some *other* person." Habermas goes on to suggest that birth, "being a natural fact, meets the conceptual requirement of constituting a beginning we cannot control. Philosophy has but rarely addressed this matter." An exception, he observes, is found in the work of Hannah Arendt, who sees "natality," the fact that human beings are born not made, as a condition of their capacity to initiate action.[35]

Habermas is onto something important, I think, when he asserts a "connection between the contingency of a life's beginning that is not at our disposal and the freedom to give one's life an ethical shape."[36] For him, this connection matters because it explains why a genetically designed child is beholden and subordinate to another person (the designing parent) in a way that a child born of a contingent, impersonal beginning is not.[37] But the notion that our freedom is bound up with "a beginning we cannot control" also carries a broader significance: Whatever its effect on the autonomy of the child, the drive to banish contingency and to master the mystery of birth diminishes the designing parent and corrupts parenting as a social practice governed by norms of unconditional love.

This takes us back to the notion of giftedness. Even if it does not harm the child or impair its autonomy, eugenic parenting is objectionable because it expresses and entrenches a certain stance toward the world—a stance of mastery and dominion that fails to appreciate the gifted character of human powers and achievements, and misses the part of freedom that consists in a persisting negotiation with the given.

5

Mastery and Gift

THE PROBLEM WITH EUGENICS and genetic engineering is that they represent the one-sided triumph of willfulness over giftedness, of dominion over reverence, of molding over beholding. But why, we may wonder, should we worry about this triumph? Why not shake off our unease with enhancement as so much superstition? What would be lost if biotechnology dissolved our sense of giftedness?

HUMILITY, RESPONSIBILITY, AND SOLIDARITY

From the standpoint of religion, the answer is clear: To believe that our talents and powers are wholly our own doing is to misunderstand our place in creation, to confuse our role with God's. But religion is not the only source of reasons to care about giftedness. The moral stakes can also be described in secular terms. If the genetic revolution erodes our appreciation for the gifted character of human powers and achievements, it will transform three key features of our moral landscape—humility, responsibility, and solidarity.

In a social world that prizes mastery and control, parenthood is a school for humility. That we care deeply about our children, and yet cannot choose the kind we want, teaches parents to be open to the unbidden. Such openness is a disposition worth affirming, not only within families but in the wider world as well. It invites us to abide the unexpected, to live with dissonance, to reign in the impulse to control. A *Gattaca*-like world, in which parents became accustomed to specifying the sex and genetic traits of their children, would be a world inhospitable to the unbidden, a gated community writ large.

The social basis of humility would also be diminished if people became accustomed to genetic self-improvement. The awareness that our talents and abilities are not wholly our own doing restrains our tendency toward hubris. If bioengineering made the myth of the "self-made man" come true, it would be difficult to view our talents as gifts for which we are indebted rather than achievements for which we are responsible. (Genetically enhanced children would of course remain indebted rather than responsible for their traits, though their debt would run more to their parents and less to nature, chance, or God.)

It is sometimes thought that genetic enhancement erodes human responsibility by overriding effort and striving. But the real problem is the explosion, not the erosion, of responsibility. As humility gives way, responsibility expands to daunting proportions. We attribute less to chance and more to choice. Parents become responsible for choosing, or failing to choose, the right traits for their children. Athletes become responsible for acquiring, or failing to acquire, the talents that will help their team win.

One of the blessings of seeing ourselves as creatures of nature, God, or fortune is that we are not wholly responsible for the way we are. The more we become masters of our genetic endowments, the greater the burden we bear for the talents we have and the way we perform. Today when a basketball player misses a rebound, his coach can blame him for being out of position. Tomorrow the coach may blame him for being too short.

Even now, the growing use of performance-enhancing drugs in professional sports is subtly transforming the expectations players have for one another. In the past when a starting pitcher's team scored too few runs to win, he could only curse his bad luck and take it in stride. These days, the use of amphetamines and other stimulants is so widespread that players who take the field without them are criticized for "playing naked." A recently retired major league outfielder told *Sports Illustrated* that some pitchers blame teammates who play unenhanced: "If the starting pitcher knows that you're going out there naked, he's upset that you're not giving him [everything] you can. The big-time pitcher wants to make sure you're beaning up before the game."[1]

The explosion of responsibility, and the moral burdens it creates, can also be seen in changing norms that accompany the use of prenatal genetic testing. Once, giving birth to a child with Down syndrome was considered a matter of chance; today many parents of children with Down syndrome or other genetic disabilities feel judged or blamed.[2] A domain once governed by fate has now become an arena of choice. Whatever one believes about which, if any, genetic conditions warrant terminating a pregnancy (or selecting against an embryo, in the case of preimplantation genetic diagnosis), the advent of genetic testing creates a burden of decision that did not exist before. Prospective parents remain free to choose whether to use prenatal testing and whether to act on the results. But they are not free to escape the burden of choice that the new technology creates. Nor can they avoid being implicated in the enlarged frame of moral responsibility that accompanies new habits of control.

The Promethean impulse is contagious. In parenting as in sports, it unsettles and erodes the gifted dimension of human experience. When performance-enhancing drugs become commonplace, unenhanced ballplayers find themselves "playing naked." When genetic screening becomes a routine part of pregnancy, parents who eschew it are regarded as "flying blind" and are held responsible for whatever genetic defect befalls their child.

Paradoxically, the explosion of responsibility for our own fate, and that of our children, may diminish our sense of solidarity with those less fortunate than ourselves. The more alive we are to the chanced nature of our lot, the more reason we have to share our fate with others. Consider the case of insurance. Since people do not know whether or when various ills will befall them, they pool their risk by buying health insurance and life insurance. As life plays itself out, the healthy wind up subsidizing the unhealthy, and those who live to a ripe old age wind up subsidizing the families of those who die before their time. The result is mutuality by inadvertence. Even without a sense of mutual obligation, people pool their risks and resources, and share one another's fate.

But insurance markets mimic the practice of solidarity only insofar as people do not know or control their own risk factors. Suppose genetic testing advanced to the point where it could reliably predict each person's medical history and life expectancy. Those confident of good health and long life would opt out of the pool, causing premiums to skyrocket for those destined for ill health. The solidaristic aspect of insurance would disappear as those with good genes fled the actuarial company of those with bad ones.

The concern that insurance companies would use genetic data to assess risks and set premiums recently led the U.S. Senate to vote to prohibit genetic discrimination in health insurance.[3] But the bigger danger, admittedly more speculative, is that genetic enhancement, if routinely practiced, would make it harder to foster the moral sentiments that social solidarity requires.

Why, after all, do the successful owe anything to the least advantaged members of society? One compelling answer to this question leans heavily on the notion of giftedness. The natural talents that enable the successful to flourish are not their own doing but, rather, their good fortune—a result of the genetic lottery.[4] If our genetic endowments are gifts, rather than achievements for which we can claim credit, it is a mistake and a conceit to assume that we are entitled to the full measure of the bounty they reap in a market economy. We therefore have an obligation to share this bounty with those who, through no fault of their own, lack comparable gifts.

Here, then, is the connection between solidarity and giftedness: A lively sense of the contingency of our gifts—an awareness that none of us is wholly responsible for his or her success—saves a meritocratic society from sliding into the smug assumption that success is the crown of virtue, that the rich are rich because they are more deserving than the poor.

If genetic engineering enabled us to override

the results of the genetic lottery, to replace chance with choice, the gifted character of human powers and achievements would recede, and with it, perhaps, our capacity to see ourselves as sharing a common fate. The successful would become even more likely than they are now to view themselves as self-made and self-sufficient, and hence wholly responsible for their success. Those at the bottom of society would be viewed not as disadvantaged, and so worthy of a measure of compensation, but as simply unfit, and so worthy of eugenic repair. The meritocracy, less chastened by chance, would become harder, less forgiving. As perfect genetic knowledge would end the simulacrum of solidarity in insurance markets, perfect genetic control would erode the actual solidarity that arises when men and women reflect on the contingency of their talents and fortunes.

OBJECTIONS

My argument against enhancement is likely to invite at least two objections: Some may complain that it is overly religious; others may object that it is unpersuasive in consequentialist terms. The first objection asserts that to speak of a gift presupposes a giver. If this is true, then my case against genetic engineering and enhancement is inescapably religious.[5] I argue, to the contrary, that an appreciation for the giftedness of life can arise from either religious or secular sources. While some believe that God is the source of the gift of life, and that reverence for life is a form of gratitude to God, one need not hold this belief in order to appreciate life as a gift or to have reverence for it. We commonly speak of an athlete's gift, or a musician's, without making any assumption about whether or not the gift comes from God. What we mean is simply that the talent in question is not wholly the athlete's or the musician's own doing; whether he has nature, fortune, or God to thank for it, the talent is an endowment that exceeds his control.

In a similar way, people often speak of the sanctity of life, and even of nature, without necessarily embracing the strong metaphysical version of that idea. For example, some hold with the ancients that nature is sacred in the sense of being enchanted, or inscribed with inherent meaning, or animated by divine purpose; others, in the Judeo-Christian tradition, view the sanctity of nature as deriving from God's creation of the universe; and still others believe that nature is sacred simply in the sense that it is not a mere object at our disposal, open to any use we may desire. These various understandings of the sacred all insist that we value nature and the living beings within it as more than mere instruments; to act otherwise displays a lack of reverence, a failure of respect. But this moral mandate need not rest on a single religious or metaphysical background.

It might be replied that nontheological notions of sanctity and gift cannot ultimately stand on their own but must lean on borrowed metaphysical assumptions they fail to acknowledge. This is a deep and difficult question that I cannot attempt to resolve here.[6] It is worth noting, however, that liberal thinkers from Locke to Kant to Habermas accept the idea that freedom depends on an origin or standpoint that exceeds our control. For Locke, our life and liberty, being inalienable rights, are not ours to give away (through suicide or selling ourselves into slavery). For Kant, though we are the authors of the moral law, we are not at liberty to exploit ourselves or to treat ourselves as objects any more than we may do so to other persons. And for Habermas, as we have seen, our freedom as equal moral beings depends on having an origin beyond human manipulation or control. We can make sense of these notions of inalienable and inviolable rights without necessarily embracing religious conceptions of the sanctity of human life. In a similar way, we can make sense of the notion of giftedness, and feel its moral weight, whether or not we trace the source of the gift to God.

The second objection construes my case against enhancement as narrowly consequentialist, and finds it wanting, along the following lines: Pointing to the possible effects of bioengineering on humility, responsibility, and solidarity may be persuasive to those who prize those virtues. But those who care more about gaining a competitive edge for their children or themselves may decide that the benefits to be gained from genetic enhancement outweigh its allegedly adverse effects on social institutions and moral sentiments. Moreover, even assuming that the desire for mastery is bad, an individual who pursues it may achieve some redeeming moral good—a cure for cancer, for example. So why should we assume that the "bad" of mastery necessarily outweighs the good it can bring about?[7]

To this objection I reply that I do not mean to rest the case against enhancement on consequentialist considerations, at least not in the usual sense of the term. My point is not that genetic engineering is objectionable simply because the social costs are likely to outweigh the benefits. Nor do I claim that people who bioengineer their children or themselves are necessarily motivated by a desire for mastery, and that this motive is a sin no good result could possibly outweigh. I am suggesting instead that the moral stakes in the enhancement debate are not fully captured by the familiar categories of autonomy and rights, on the one hand, and the calculation of costs and benefits, on the other. My concern with enhancement is not as individual vice but as habit of mind and way of being.[8]

The bigger stakes are of two kinds. One involves the fate of human goods embodied in important social practices—norms of unconditional love and an openness to the unbidden, in the case of parenting; the celebration of natural talents and gifts in athletic and artistic endeavors; humility in the face of privilege, and a willingness to share the fruits of good fortune through institutions of social solidarity. The other involves our orientation to the world that we inhabit, and the kind of freedom to which we aspire.

It is tempting to think that bioengineering our children and ourselves for success in a competitive society is an exercise of freedom. But changing our nature to fit the world, rather than the other way around, is actually the deepest form of disempowerment. It distracts us from reflecting critically on the world, and deadens the impulse to social and political improvement. Rather than employ our new genetic powers to straighten "the crooked timber of humanity,"[9] we should do what we can to create social and political arrangements more hospitable to the gifts and limitations of imperfect human beings.

The Project of Mastery

In the late 1960s, Robert L. Sinsheimer, a molecular biologist at the California Institute of Technology, glimpsed the shape of things to come. In an article entitled "The Prospect of Designed Genetic Change," he argued that freedom of choice would vindicate the new genetics, and set it apart from the discredited eugenics of old. "To implement the older eugenics of Galton and his successors would have required a massive social program carried out over many generations. Such a program could not have been initiated without the consent and cooperation of a major fraction of the population, and would have been continuously subject to social control. In contrast, the new eugenics could, at least in principle, be implemented on a quite individual basis, in one generation, and subject to no existing restrictions."[10]

According to Sinsheimer, the new eugenics would be voluntary rather than coerced, and also more humane. Rather than segregate and eliminate the unfit, it would improve them. "The old eugenics would have required a continual selection for breeding of the fit, and a culling of the unfit. The new eugenics would permit in principle the conversion of all the unfit to the highest genetic level."[11]

Sinsheimer's paean to genetic engineering caught the heady, Promethean self-image of the age. He wrote hopefully of rescuing "the losers in that chromosomal lottery that so firmly channels our human destinies," including not only those born with genetic defects but also "the 50 million 'normal' Americans with an IQ of less than 90." But he also saw that something bigger was at stake than improving upon nature's "mindless, age-old throw of dice." Implicit in the new technologies of genetic intervention was a new, more exalted place for human beings in the cosmos. "As we enlarge man's freedom, we diminish his constraints and that which he must accept as given." Copernicus and Darwin had "demoted man from his bright glory at the focal point of the universe," but the new biology would restore his pivotal role. In the mirror of our new genetic knowledge, we would see ourselves as more than a link in the chain of evolution: "We can be the agent of transition to a whole new pitch of evolution. This is a cosmic event."[12]

There is something appealing, even intoxicating, about a vision of human freedom unfettered by the given. It may even be the case that the allure of that vision played a part in summoning the genomic age into being. It is often assumed that the powers of enhancement we now possess arose as an inadvertent by-product of biomedical progress—the genetic revolution came, so to speak, to cure disease, but stayed to tempt us with the prospect of enhancing our performance, designing our children, and perfecting our nature. But that may have the story backward. It is also possible to view genetic engineering as the ultimate expression of our resolve to see ourselves astride the world, the masters of our nature. But that vision of freedom is flawed. It threatens to banish our appreciation of life as a gift, and to leave us with nothing to affirm or behold outside our own will.

EPILOGUE

Embryo Ethics: The Stem Cell Debate

IN OPPOSING GENETIC ENHANCEMENT, I
have argued against the one-sided triumph of mastery over reverence, and have urged that we re-
claim an appreciation of life as a gift. But I have
also argued that there is a difference between healing and enhancing. Medicine intervenes in nature,
but because it is constrained by the goal of restoring normal human functioning, it does not represent an unbridled act of hubris or bid for dominion. The need for healing arises from the fact that
the world is not perfect and complete but in con-
stant need of human intervention and repair. Not
everything given is good. Smallpox and malaria are
not gifts, and it would be good to eradicate them.

The same can be said of diabetes, Parkinson's
disease, ALS, and spinal cord injuries. One of the
most promising new sources of hope for people
afflicted with these conditions is stem cell research.

Scientists may soon be able to extract stem cells from an early embryo and grow those cells to study and cure degenerative diseases. Critics object that extracting the stem cells destroys the embryo. They argue that if life is a gift, then research that destroys nascent human life must surely be rejected. In this chapter, I offer a defense of embryonic stem cell research and try to show that the ethic of giftedness does not condemn it.

STEM CELL QUESTIONS

In the summer of 2006, well into the sixth year of his presidency, George W. Bush exercised his first veto. The bill he rejected involved not a familiar Washington issue like taxes or terrorism or the war in Iraq, but the more arcane subject of stem cell research. Hoping to promote cures for diabetes, Parkinson's, and other degenerative diseases, Congress had voted to fund new embryonic stem cell research, in which scientists isolate cells capable of becoming any tissue in the body. The President refused to go along. He argued that the research is unethical because deriving these cells destroys the blastocyst, an unimplanted embryo at the sixth to eighth day of development. The federal government, he declared, should not support "the taking of innocent human life."[1]

The President's press secretary could be forgiven his confusion. In explaining the veto, he stated that the President considered embryonic stem cell research to be "murder," something the federal government should not support. When the comment drew a flurry of critical press attention, the White House retreated. No, the President did not believe that destroying an embryo was murder. The press secretary retracted his statement, and apologized for having "overstated the President's position."[2]

How exactly the spokesman had overstated the President's position is unclear. If embryonic stem cell research does constitute the deliberate taking of innocent human life, it is hard to see how it differs from murder. The chastened press secretary made no attempt to parse the distinction. He was not the first to become entangled in the ethical and political complexities of the stem cell debate.

The debate over stem cell research poses three questions. First, should embryonic stem cell research be permitted? Second, should it be funded by the government? Third, should it matter, for either permissibility or funding, whether the stem cells are taken from already existing embryos left over from fertility treatments or from cloned embryos created for research?

The first question is the most fundamental and, some would say, the most intractable. The main objection to embryonic stem cell research is that destroying a human embryo, even in its earliest stages of development, and even for the sake of noble ends, is morally abhorrent; it is like killing a child to save other people's lives. The validity of this objection depends, of course, on the moral status of the embryo. Since some people hold strong religious convictions on the question, it is sometimes thought that it is not subject to rational argument or analysis. But that is a mistake. The fact that a moral belief may be rooted in religious conviction neither exempts it from challenge nor renders it incapable of rational defense.

Later in this chapter I will try to show how moral reasoning about the status of the embryo can proceed. But to prepare the way, I turn first to the question of whether there is a moral difference between the use of "spare" or "excess" embryos left over from fertility treatments and the use of cloned embryos created for research. Many politicians believe that there is.

CLONES AND SPARES

To this day, the United States has no federal law that prohibits cloning a child. This is not because most people favor cloning as a new means of reproduction. To the contrary, public opinion and almost all elected officials oppose it.

But there is strong disagreement about whether to permit cloning to create embryos for stem cell research. And the opponents of research cloning have so far been unwilling to support a separate ban on reproductive cloning, as Britain has enacted.[3] In 2001, the House of Representatives passed a bill that would have banned not only reproductive cloning but also cloning for biomedical research. The bill did not become law because Senate supporters of stem cell research were unwilling to accept the blanket ban. As a result of this stalemate, the United States has no federal law against human reproductive cloning.

The debate over cloning brought out two different reasons for opposing the use of cloned embryos in stem cell research. Some people oppose research cloning on the grounds that the embryo is a person. They maintain that all embryonic stem cell research is immoral (whether on cloned or natural embryos), because it amounts to killing a person to treat other people's diseases. This is the position of Senator Sam Brownback of Kansas, a leading advocate of the right-to-life position. Embryonic stem cell research is wrong, he argues, because "it is never acceptable to deliberately kill one innocent human being in order to help another."[4] If the embryo is a person, then harvesting its stem cells is morally analogous to harvesting organs from babies. In Brownback's view, "A human embryo . . . is a human being just like you and me; and it deserves the same respect that our laws give to us all."[5]

Other opponents of research cloning do not go that far. They support embryonic stem cell research, provided it uses "spare" embryos left over from fertility clinics.[6] They are troubled by the deliberate creation of embryos for research. But since in vitro fertilization clinics create many more fertilized eggs than are ultimately implanted, some people argue that there is nothing wrong with using those spares for research. If the excess embryos would be discarded anyway, they reason, why not use them (with donor consent) for potentially lifesaving research?

To politicians looking for a principled compromise in the stem cell debate, this position holds considerable appeal. Since it would sanction the use only of excess embryos, it would seem to overcome moral qualms about creating embryos for the sake of research. This position was defended in the Senate by Majority Leader Bill Frist of Tennessee, the Senate's only physician, and in Massachusetts by Governor Mitt Romney, who unsuccessfully urged his legislature to adopt it. Both supported stem cell research on leftover embryos created for reproduction, but not on embryos created for research.[7] The stem cell funding bill voted by Congress (and vetoed by President Bush) in 2006 also made this distinction; it would have funded stem cell research only on embryos left over from fertility treatments.

Beyond its appeal as a political compromise, this distinction seems morally defensible as well. On closer examination, however, it does not hold up. The distinction fails because it begs the question of whether the "spare" embryos should be created in the first place. To see how this is so, imagine a fertility clinic that accepts egg and sperm donations for two purposes—reproduction and stem cell research. No cloning is involved. The clinic creates two groups of embryos, one from eggs and sperm donated for the purpose of IVF, the other from eggs and sperm donated by people who want to advance the cause of stem cell research.

Which group of embryos may an ethical scientist use for stem cell research? Those who agree with Frist and Romney are left in a paradoxical position: They would permit the scientist to use spare embryos from the first group (since they were created for reproduction and will otherwise be discarded) but not from the second group (since they were deliberately created for research). In fact, Frist and Romney have both sought to ban the deliberate creation of embryos in IVF clinics for purposes of research.

The paradoxical scenario brings out the flaw in the compromise position: Those who oppose the creation of embryos for stem cell research but support research on IVF "spares" fail to address the morality of in vitro fertilization itself. If it is immoral to create and sacrifice embryos for the sake of curing or treating devastating diseases, why isn't it also objectionable to create and discard spare embryos in the course of treating infertility? Or, to look at the argument from the opposite end, if the creation and sacrifice of embryos in IVF is morally acceptable, why isn't the creation and sacrifice of embryos for stem cell research also acceptable? After all, both practices serve worthy ends, and curing diseases such as Parkinson's and diabetes is at least as important as treating infertility.

Those who see a moral difference between the sacrifice of embryos in IVF and the sacrifice of embryos in stem cell research might reply as follows: The fertility doctor who creates excess embryos does so to increase the odds of a successful pregnancy; he does not know which embryos will ultimately be discarded, and does not intend the death of any. But the scientist who deliberately creates an embryo for stem cell research knows the embryo will die, for to carry out the research it is necessary to destroy the embryo. Charles Krauthammer, who favors stem cell research on IVF spares but not on embryos created for research, put the point sharply: "The bill that would legalize research cloning essentially sanctions . . . a most ghoulish enterprise: the creation of nascent human life for the sole purpose of its exploitation and destruction."[8]

This reply is unpersuasive, for two reasons. First, the claim that creating embryos for stem cell research amounts to creating life *for the purpose* of exploiting or destroying it is misleading. The destruction of the embryo is, admittedly, a foreseeable consequence of the act, but the purpose is curing disease. Those who create embryos for research no more aim at destruction or exploitation than those who create embryos for fertility treatments aim at discarding spares.[9]

Second, although fertility doctors and patients do not know in advance which of the embryos they create will wind up being discarded, the fact remains that IVF, as practiced in the United States, generates tens of thousands of excess embryos bound for destruction. (A recent study found that some 400,000 frozen embryos are languishing in American fertility clinics, with another 52,000 in the United Kingdom and 71,000 in Australia.)[10] It is true that, once these doomed embryos exist, "nothing is lost" if they are used for research.[11] But whether they should be created in the first place is as much a policy choice as whether to permit the creation of embryos for research. German federal law, for example, regulates fertility clinics and prohibits doctors from fertilizing more eggs than will be implanted at any one time. As a result, German IVF clinics do not generate excess embryos. The existence of large numbers of doomed embryos in the freezers of U.S. fertility clinics is not an unalterable fact of nature but the consequence of a policy that elected officials could change if they wanted to. So far, however, few of those who would ban the creation of embryos for research have called for a ban on the creation and destruction of excess embryos in fertility clinics.

Whoever is right about the moral status of the embryo, one thing is clear: Opponents of research cloning cannot have it both ways. They cannot endorse the creation and destruction of excess embryos in fertility clinics, or the use of such embryos in research, and at the same time complain that creating embryos for research and regenerative medicine is morally objectionable. If cloning for stem cell research violates the respect the embryo is due, then so does stem cell research on IVF spares, and so does any fertility treatment that creates and discards excess embryos.

Those, like Senator Brownback, who take a consistent stand against the use of embryonic human life are right at least to this extent: The moral arguments for research cloning and for stem cell research on leftover embryos stand or fall together. It remains to ask whether they stand or fall. This takes us to the basic question of whether any embryonic stem cell research should be permitted.

THE MORAL STATUS OF THE EMBRYO

There are two main arguments against permitting embryonic stem cell research. One holds that, despite its worthy ends, stem cell research is wrong because it involves the destruction of human embryos; the other worries that even if research on embryos is not wrong in itself, it would open the way to a slippery slope of dehumanizing practices, such as embryo farms, cloned babies, the use of fetuses for spare parts, and the commodification of human life.

The slippery slope objection is a practical one that deserves to be taken seriously. But its worries could be addressed by adopting regulatory safeguards to prevent embryo research from devolving into nightmare scenarios of exploitation and abuse. The first objection, however, is more philosophically challenging. Whether it is decisive depends on whether its view of the moral status of the embryo is correct.

It is important to be clear, first of all, about the embryo from which stem cells are extracted. It is not a fetus. It has no recognizable human features or form. It is not an embryo implanted and growing in a woman's uterus. It is, rather, a blastocyst, a cluster of 180 to 200 cells, growing in a petri dish, barely visible to the naked eye. The blastocyst represents such an early stage of embryonic development that the cells it contains have not yet differentiated, or taken on the properties of particular organs or tissues—kidneys, muscles, spinal cord, and so on. This is why the stem cells that are extracted from the blastocyst hold the promise of developing, with proper coaxing in the lab, into any kind of cell the researcher wants to study or repair. The moral and political controversy arises from the fact that extracting the stem cells destroys the blastocyst.

To assess this controversy, one must begin by grasping the full force of the claim that the embryo is morally equivalent to a person, a fully developed human being. For those who hold this view, extracting stem cells from a blastocyst is as morally abhorrent as harvesting organs from a baby to save other people's lives. Some base this claim on the religious belief that ensoulment occurs at conception. Others defend it without recourse to religion, by the following line of reasoning:

> Human beings are not things; their lives must not be sacrificed against their will, even for the sake

of good ends, like saving other people's lives. The reason human beings must not be treated as things or used merely as means to an end is that they are inviolable. They are, to borrow Kant's language, ends in themselves, worthy of respect. At what point do we acquire this inviolability? When does human life become worthy of respect? The answer cannot depend on the age or developmental stage of a particular human life. Infants are clearly inviolable, and few people would countenance harvesting organs for transplantation even from a fetus. Every human being—each one of us—began life as an embryo. If our lives are worthy of respect, and hence inviolable, simply by virtue of our humanity, it would be a mistake to think that at some younger age or earlier stage of development we were not worthy of respect. Unless we can point to a definitive moment in the passage from conception to birth that marks the emergence of the human person, we must regard embryos as possessing the same inviolability as fully developed human beings.

I will try to show that this argument is unpersuasive on two levels: Its reasoning is flawed, and it carries moral implications that even its defenders find difficult to embrace. Before turning to these difficulties, however, I want to acknowledge the validity of two aspects of the equal-moral-status position. First, it rightly rejects the utilitarian view of morality, which weighs costs and benefits without regard for the inviolability of persons. Second, it is undeniable that the blastocyst is "human life," at least in the obvious sense that it is living rather than dead, and human rather than, say, bovine. But it does not follow from this biological fact that the blastocyst is a human being, or a person. Any living human cell (a skin cell, for example) is "human life" in the sense of being human rather than bovine and living rather than dead. But no one would consider a skin cell a person, or deem it inviolable. Showing that a blastocyst is a human being, or a person, requires further argument.

Analyzing the Argument

The argument for the equal-moral-status view begins with the observations that every person was once an embryo, and that there is no nonarbitrary line between conception and adulthood that can tell us when personhood begins. It then asserts that, given the lack of such a line, we should regard the blastocyst as a person, morally equivalent to a fully developed human being. But this argument is not persuasive, for several reasons.[12]

First, a small but not inconsequential point: While it is true that every one of us was once an embryo, none of us was ever a cloned blastocyst. So even if the fact of our embryonic origin did prove that embryos are persons, it would only condemn stem cell research on embryos produced by the union of egg and sperm, not stem cell research on cloned embryos. In fact, some participants in the stem cell debate have argued that cloned blastocysts are not, strictly speaking, embryos but biologic artifacts ("clonotes" rather than zygotes) that lack the moral status of naturally conceived human embryos. They argue that using cloned embryos for research is thus morally less troubling than using natural ones.[13]

Second, even setting aside the question of the "clonote," the fact that every person began life as an embryo does not prove that embryos are persons. Consider an analogy: Although every oak tree was once an acorn, it does not follow that acorns are oak trees, or that I should treat the loss of an acorn eaten by a squirrel in my front yard as the same kind of loss as the death of an oak tree felled by a storm.[14] Despite their developmental continuity, acorns and oak trees differ. So do human embryos and human beings, and in the same way. Just as acorns are potential oaks, human embryos are potential human beings. The distinction between actual persons and potential ones is not without ethical significance. Sentient creatures make claims on us that nonsentient ones do not; beings capable of experience and consciousness make higher claims still. Human life develops by degrees.

Defenders of the equal-moral-status view challenge their interlocutors to specify a nonarbitrary moment in the course of human development when personhood, or inviolability, sets in. If the embryo is not a person, then when exactly do we become persons? This is not a question that admits an easy answer. Many people point to birth as the moment that marks the advent of personhood. But this answer is open to the objection that it would surely be wrong to dismember a late-stage human fetus for the sake of medical research. (Beyond inviolability, there are other aspects of personhood—having a name, for example—that unfold, depending on the culture or tradition, at various times after birth.)

The difficulty of specifying the exact beginning of personhood along the developmental continuum does not establish, however, that blastocysts are persons. Consider an analogy: Suppose someone asked you how many grains of wheat constitute a heap? One grain does not, nor two, nor three. The fact that there is no nonarbitrary point when the addition of one more grain will bring a heap into being does not mean that there is no difference between a grain and a heap. Nor does it give us reason to conclude that a grain must be a heap.

This puzzle about specifying points along a continuum, known to philosophers as the "sorites paradox," goes back to the ancient Greeks. ("Sorites" comes from *soros*, the Greek word for "heap.") The sophists used sorites arguments to try to persuade their listeners that two separate qualities linked by a continuum were actually the same, even if intuition and common sense suggested otherwise.[15] Baldness is a classic example. Everyone would agree that a man with only one hair on his head is bald. What number of hairs marks the transition from being bald to having a full head of hair? Although there is no determinate answer to this question, it does not follow that there is no difference between being bald and having a full head of hair. The same is true of human personhood. The fact of developmental continuity from blastocyst to implanted embryo to fetus to newborn child does not establish that a baby and a blastocyst are, morally speaking, one and the same.

The arguments from embryonic origin and developmental continuity thus do not compel the conclusion that the blastocyst is inviolable, the moral equivalent of a person. Beyond identifying the flaws in its reasoning, one can challenge the equal-moral-status position from another standpoint as well. Perhaps the best way to see its implausibility is to notice that even those who invoke it hesitate to embrace its full implications.

Pursuing the Implications

In 2001 President Bush announced a policy that restricted federal funding to already existing stem cell lines, so that no taxpayer funds would encourage or support the destruction of embryos. And in 2006 he vetoed a bill that would have funded new embryonic stem cell research, saying that he did not want to support "the taking of innocent human life." But it is a striking feature of the President's position that, while restricting the funding of embryonic stem cell research, he has made no effort to ban it. To adapt a slogan from an earlier president's quandary, the Bush policy might be summarized as

"Don't Fund, Don't Ban." But this policy fits uneasily with the notion that the embryo is a human being.

If harvesting stem cells from a blastocyst were truly on a par with harvesting organs from a baby, then the morally responsible policy would be to ban it, not merely deny it federal funding. If some doctors made a practice of killing children to get organs for transplantation, no one would take the position that the infanticide should be ineligible for federal funding but allowed to continue in the private sector. In fact, if we were persuaded that embryonic stem cell research were tantamount to infanticide, we would not only ban it but treat it as a grisly form of murder and subject scientists who performed it to criminal punishment.

It might be argued, in defense of the President's policy, that Congress would be unlikely to enact an outright ban on embryonic stem cell research. But this does not explain why, if the President really considers embryos to be human beings, he has not at least called for such a ban, nor even called on scientists to stop doing stem cell research that involves the destruction of embryos. To the contrary, President Bush has cited the fact that "there is no ban on embryonic stem cell research" in touting the virtues of his "balanced approach."[16]

The moral oddness of the Bush "Don't Fund, Don't Ban" position makes his press secretary's gaffe entirely understandable. The spokesman's errant statement that the President considered embryo destruction to be "murder" simply followed the moral logic of the notion that embryos are human beings. It was a gaffe only because the Bush policy did not follow the full implications of that logic.

Defenders of the equal-moral-status view might simply reply that they part company with politicians who shrink from pursuing the full implications of their position, whether by failing to ban embryonic stem cell research or failing to ban fertility treatments that create and discard excess embryos. Even the most principled politicians compromise their principles from time to time; this is hardly unique to those who profess the belief that embryos are human beings. But even putting politics aside, principled advocates of the equal-moral-status view might be hard-pressed to endorse the full implications of their position.

Consider the following hypothetical (first suggested, so far as I know, by George Annas)[17]: Suppose a fire broke out in a fertility clinic, and you had time to save either a five-year-old girl or a tray of twenty frozen embryos. Would it be wrong to save the girl? I have yet to encounter a proponent of the equal-moral-status view who is willing to say that he or she would rescue the tray of embryos. But if you really believed that those embryos were human beings, and all other things were equal (that is, you had no personal connection to either the girl or the embryos), on what possible grounds could you justify saving the girl?

Or consider a less hypothetical case. I recently participated in a stem cell debate with a proponent of the view that a blastocyst is morally equivalent to a baby. After our exchange, a member of the audience related a personal experience. He and his wife had successfully conceived three children by means of in vitro fertilization. They had no desire for more children, and yet three viable embryos remained. What, he asked, should he and his wife do with these excess embryos?

My right-to-life interlocutor replied that it would be wrong to exploit the embryos by using them (and destroying them) for stem cell research. Assuming no one was available to adopt them, the only thing to do was to let them die with dignity. Given the assumption that these embryos were morally equivalent to children, I could not quarrel with his conclusion. If we encountered some prisoners unjustly doomed to death, it would not be right to say, "We may as well make the best of a bad situation and extract their organs for transplantation."

What I found puzzling about his answer was not his unwillingness to sanction the use of the embryos for research, but his reluctance to articulate the full implications of his position. If those embryos really are young human beings, then the honest answer would be to tell the questioner that what he and his wife did in creating and discarding those embryos was nothing less than creating three surplus siblings of their children, and then abandoning the unwanted siblings to die by exposure on a mountainside (or in a freezer). But if that description is morally apt—if the 400,000 excess embryos frozen in U.S. fertility clinics are like newborns left to die on a mountainside—then why are opponents of embryonic stem cell research not leading a campaign to shut down what they must regard as rampant infanticide?

Those who consider embryos to be persons

might reply that they do indeed oppose fertility treatments that create and discard excess embryos, but that they have little hope of banning the practice. But the full implications of their position point beyond even a concern for embryos lost in IVF. Defenders of in vitro fertilization point out that the rate of embryo loss in assisted reproduction is actually less than in natural pregnancy, in which more than half of all fertilized eggs either fail to implant or are otherwise lost. This fact highlights a further difficulty with the view that equates embryos and persons. If early embryo death is a common occurrence in natural procreation, perhaps we should worry less about the loss of embryos that occurs in fertility treatments and stem cell research.[18]

Those who view embryos as persons reply, rightly, that a high rate of infant mortality would not justify infanticide. But the way we respond to the natural loss of embryos suggests that we do not regard this event as the moral or religious equivalent of the death of an infant. Even those religious traditions that are the most solicitous of nascent human life do not mandate the same burial rituals for the loss of an embryo as for the death of a child. Moreover, if the embryo loss that accompanies natural procreation were the moral equivalent of infant death, then pregnancy would have to be regarded as a public health crisis of epidemic proportions; alleviating natural embryo loss would be a more urgent moral cause than abortion, in vitro fertilization, and stem cell research combined. But few who are stirred by these familiar causes are mounting ambitious campaigns or seeking new technologies to prevent or reduce embryo loss in natural pregnancy.

The Warrant for Respect

Having criticized the view that regards embryos as human beings, I do not suggest that embryos are mere things, open to any use we may desire or devise. Embryos are not inviolable, but neither are they objects at our disposal. Those who view embryos as persons often assume that the only alternative is to treat them with moral indifference. But one need not regard the embryo as a full human being to accord it a certain respect. To regard an embryo as a mere thing misses its significance as potential human life. Few would sanction the wanton destruction of embryos or the use of embryos for the purpose of developing a new line of cosmetics. But the notion that human embryos should not be treated as mere objects does not prove that they are persons.

Personhood is not the only warrant for respect. If an eccentric billionaire bought van Gogh's *Starry Night* and used it as a doormat, such use would be a kind of sacrilege, a scandalous failure of respect—not because we regard the painting as a person but because, as a great work of art, it is worthy of a higher mode of valuation than mere use. We also consider it an act of disrespect when a thoughtless hiker carves his initials in an ancient sequoia—not because we regard the sequoia as a person but because we regard it as a natural wonder worthy of appreciation and awe. To respect the old-growth forest does not mean that no tree may ever be felled or harvested for human purposes. Respecting the forest may be consistent with using it. But the purposes should be weighty and appropriate to the wondrous nature of the thing.

The conviction that the embryo is a person derives support not only from certain religious doctrines but also from the Kantian assumption that the moral universe is divided in binary terms: everything is either a person, worthy of respect, or a thing, open to use. But as the van Gogh and sequoia examples suggest, this dualism is overdrawn.

The way to combat the instrumentalizing tendencies of modern technology and commerce is not to insist on an all-or-nothing ethic of respect for persons that consigns the rest of life to a utilitarian calculus. Such an ethic risks turning every moral question into a battle over the bounds of personhood. We would do better to cultivate a more expansive appreciation of life as a gift that commands our reverence and restricts our use. Genetic engineering to create designer babies is the ultimate expression of the hubris that marks the loss of reverence for life as a gift. But stem cell research to cure debilitating disease, using unimplanted blastocysts, is a noble exercise of our human ingenuity to promote healing and to play our part in repairing the given world.

Those who warn of slippery slopes, embryo farms, and the commodification of ova and zygotes are right to worry but wrong to assume that embryo research necessarily opens us to these dangers. Rather than ban embryonic stem cell research and research cloning, we should allow them to proceed subject to regulations that embody the moral restraint appropriate to the mystery of the first stirrings of human life. Such regulations should include a ban on human reproductive cloning, reasonable limits on the length of time an embryo can be grown in the lab, licensing requirements for fertility clinics, restrictions on the commodification of eggs and sperm, and a stem cell bank to prevent proprietary interests from monopolizing access to stem cell lines. This approach, it seems to me, offers the best hope of avoiding the wanton use of nascent human life and making biomedical advance a blessing for health rather than an episode in the erosion of our human sensibilities.

Notes

Index

Notes

1. THE ETHICS OF ENHANCEMENT

1. Margarette Driscoll, "Why We Chose Deafness for Our Children," *Sunday Times* (London), April 14, 2002. See also Liza Mundy, "A World of Their Own," *Washington Post*, March 31, 2002, p. W22.
2. Driscoll, "Why We Chose Deafness."
3. See Gina Kolata, "$50,000 Offered to Tall, Smart Egg Donor," *New York Times*, March 3, 1999, p. A10.
4. Alan Zarembo, "California Company Clones a Woman's Cat for $50,000," *Los Angeles Times*, December 23, 2004.
5. See Web site for Genetic Savings & Clone, at *http://www.savingsandclone.com*; Zarembo, "California Company Clones a Woman's Cat."
6. The phrase "better than well" is from Carl Elliott, *Better Than Well: American Medicine*

Meets the American Dream (New York: W. W. Norton, 2003), who in turn cites Peter D. Kramer, *Listening to Prozac*, rev. ed. (New York: Penguin, 1997).

7. E. M. Swift and Don Yaeger, "Unnatural Selection," *Sports Illustrated*, May 14, 2001, p. 86; H. Lee Sweeney, "Gene Doping," *Scientific American*, July 2004, pp. 62–69.

8. Richard Sandomir, "Olympics: Athletes May Next Seek Genetic Enhancement," *New York Times*, March 21, 2002, p. 6.

9. Rick Weiss, "Mighty Smart Mice," *Washington Post*, September 2, 1999, p. A1; Richard Saltus, "Altered Genes Produce Smart Mice, Tough Questions," *Boston Globe*, September 2, 1999, p. A1; Stephen S. Hall, "Our Memories, Our Selves," *New York Times Magazine*, February 15, 1998, p. 26.

10. Hall, "Our Memories, Our Selves," p. 26; Robert Langreth, "Viagra for the Brain," *Forbes*, February 4, 2002; David Tuller, "Race Is On for a Pill to Save the Memory," *New York Times*, July 29, 2003; Tim Tully et al., "Targeting the CREB Pathway for Memory Enhancers," *Nature* 2 (April 2003): 267–277; *www.memorypharma.com*.

11. Ellen Barry, "Pill to Ease Memory of Trauma Envisioned," *Boston Globe*, November 18, 2002,

p. A1; Robin Maranz Henig, "The Quest to Forget," *New York Times Magazine*, April 4, 2004, pp. 32–37; Gaia Vince, "Rewriting Your Past," *New Scientist*, December 3, 2005, p. 32.

12. Marc Kaufman, "FDA Approves Wider Use of Growth Hormone," *Washington Post*, July 26, 2003, p. A12.

13. Patricia Callahan and Leila Abboud, "A New Boost for Short Kids," *Wall Street Journal*, June 11, 2003.

14. Kaufman, "FDA Approves Wider Use of Growth Hormone"; Melissa Healy, "Does Shortness Need a Cure?" *Los Angeles Times*, August 11, 2003.

15. Callahan and Abboud, "A New Boost for Short Kids."

16. Talmud, *Niddah* 31b, cited in Miryam Z. Wahrman, *Brave New Judaism: When Science and Scripture Collide* (Hanover, NH: Brandeis University Press, 2002), p. 126; Meredith Wadman, "So You Want a Girl?" *Fortune*, February 19, 2001, p. 174; Karen Springen, "The Ancient Art of Making Babies," *Newsweek*, January 26, 2004, p. 51.

17. Susan Sachs, "Clinics' Pitch to Indian Emigrés: It's a Boy," *New York Times*, August 15, 2001, p. A1; Seema Sirohi, "The Vanishing Girls of In-

dia," *Christian Science Monitor,* July 30, 2001, p. 9; Mary Carmichael, "No Girls, Please," *Newsweek,* January 26, 2004; Scott Baldauf, "India's 'Girl Deficit' Deepest among Educated," *Christian Science Monitor,* January 13, 2006, p. 1; Nicholas Eberstadt, "Choosing the Sex of Children: Demographics," presentation to President's Council on Bioethics, October 17, 2002, at *www.bioethics.gov/transcripts/oct02/session2.html;* B. M. Dickens, "Can Sex Selection Be Ethically Tolerated?" *Journal of Medical Ethics* 28 (December 2002): 335–336; "Quiet Genocide: Declining Child Sex Ratios," *Statesman* (India), December 17, 2001.

18. See the Genetics & IVF Institute Web site, at *www.microsort.net;* see also Meredith Waldman, "So You Want a Girl?"; Lisa Belkin, "Getting the Girl," *New York Times Magazine,* July 25, 1999; Claudia Kalb, "Brave New Babies," *Newsweek,* January 26, 2004, pp. 45–52.

19. Felicia R. Lee, "Engineering More Sons than Daughters: Will It Tip the Scales toward War?" *New York Times,* July 3, 2004, p. B7; David Glenn, "A Dangerous Surplus of Sons?" *Chronicle of Higher Education,* April 30, 2004, p. A14; Valerie M. Hudson and Andrea M. den Boer,

Bare Branches: Security Implications of Asia's Surplus Male Population (Cambridge, MA: MIT Press, 2004).

20. See *www.microsort.net.*

2. Bionic Athletes

1. For this reason, I do not agree with the main thrust of the analysis of performance enhancement presented in *Beyond Therapy: Biotechnology and the Pursuit of Happiness*, A Report of the President's Council on Bioethics (Washington, DC: 2003), pp. 123–156, at *http://www.bioethics.gov/reports/beyondtherapy/index.html.*

2. Hank Gola, "Fore! Look Out for Lasik," *Daily News,* May 28, 2002, p. 67.

3. See Malcolm Gladwell, "Drugstore Athlete," *New Yorker,* September 10, 2001, p. 52, and Neal Bascomb, *The Perfect Mile* (London: CollinsWillow, 2004).

4. See Andrew Tilin, "The Post-Human Race," *Wired,* August 2002, pp. 82–89, 130–131, and Andrew Kramer, "Looking High and Low for Winners," *Boston Globe,* June 8, 2003.

5. See Matt Seaton and David Adam, "If This

Year's Tour de France Is 100% Clean, Then That Will Certainly Be a First," *Guardian,* July 3, 2003, p. 4, and Gladwell, "Drugstore Athlete."

6. Gina Kolata, "Live at Altitude? Sure. Sleep There? Not So Sure," *New York Times,* July 26, 2006, p. C12; Christa Case, "Athlete Tent Gives Druglike Boost. Should It Be Legal?" *Christian Science Monitor,* May 12, 2006; I am grateful to Thomas H. Murray, chairman of the ethics panel of the World Anti-Doping Agency, for providing me a copy of the panel's memo, "WADA Note on Artificially Induced Hypoxic Conditions," May 24, 2006.

7. Selena Roberts, "In the NFL, Wretched Excess Is the Way to Make the Roster," *New York Times,* August 1, 2002, pp. A21, A23.

8. Ibid., p. A23.

9. I am indebted to Leon Kass for suggesting the *Chariots of Fire* example.

10. See Blair Tindall, "Better Playing through Chemistry," *New York Times,* October 17, 2004.

11. Anthony Tommasini, "Pipe Down! We Can Hardly Hear You," *New York Times,* January 1, 2006, pp. AR1, AR25.

12. Ibid., p. AR25.

13. Ibid.

14. G. Pascal Zachary, "Steroids for Everyone!" *Wired*, April 2004.

15. *PGA Tour, Inc., v. Casey Martin*, 532 U.S. 661 (2001). Justice Scalia dissenting, at 699–701.

16. Hans Ulrich Gumbrecht makes a similar point when he describes athletic excellence as an expression of beauty worthy of praise. See Gumbrecht, *In Praise of Athletic Beauty* (Cambridge, MA: Harvard University Press, 2006). Tony LaRussa, one of baseball's greatest managers, applies the category of beauty to plays that capture the subtle essence of the game: "Beautiful. Just beautiful baseball." Quoted in Buzz Bissinger, *Three Nights in August* (Boston: Houghton Mifflin, 2005), pp. 2, 216–217, 253.

3. Designer Children, Designing Parents

1. William F. May's comments to President's Council on Bioethics, October 17, 2002, at *http://bioethicsprint.bioethics.gov/transcripts/oct02/session2.html*.

2. Julian Savulescu, "New Breeds of Humans: The Moral Obligation to Enhance," *Ethics, Law and Moral Philosophy of Reproductive Biomedicine* 1,

no. 1 (March 2005): 36–39; Julian Savulescu, "Why I Believe Parents Are Morally Obliged to Genetically Modify Their Children," *Times Higher Education Supplement*, November 5, 2004, p. 16.

3. William F. May's comments to President's Council on Bioethics, January 17, 2002, at *www.bioethics.gov/transcripts/jan02/jansession2intro.html*. See also William F. May, "The President's Council on Bioethics: My Take on Some of Its Deliberations," *Perspectives in Biology and Medicine* 48 (Spring 2005): 230–231.

4. Ibid.

5. See Alvin Rosenfeld and Nicole Wise, *Hyper-Parenting: Are You Hurting Your Child by Trying Too Hard?* (New York: St. Martin's Press, 2000).

6. Robin Finn, "Tennis: Williamses Are Buckled in and Rolling, at a Safe Pace," *New York Times*, November 14, 1999, sec. 8, p. 1; Steve Simmons, "Tennis Champs at Birth," *Toronto Sun*, August 19, 1999, p. 95.

7. Dale Russakoff, "Okay, Soccer Moms and Dads: Time Out!" *Washington Post*, August 25, 1998, p. A1; Jill Young Miller, "Parents, Behave! Soccer Moms and Dads Find Themselves Graded on Conduct, Ordered to Keep Quiet," *Atlanta Journal and Constitution*, October 9, 2000,

p. 1D; Tatsha Robertson, "Whistles Blow for Alpha Families to Call a Timeout," *Boston Globe*, November 26, 2004, p. A1.

8. Bill Pennington, "Doctors See a Big Rise in Injuries as Young Athletes Train Nonstop," *New York Times*, February 22, 2005, pp. A1, C19.

9. Tamar Lewin, "Parents' Role Is Narrowing Generation Gap on Campus," *New York Times*, January 6, 2003, p. A1.

10. Jenna Russell, "Fending Off the Parents," *Boston Globe*, November 20, 2002, p. A1; see also Marilee Jones, "Parents Get Too Aggressive on Admissions," *USA Today*, January 6, 2003, p. 13A; Barbara Fitzgerald, "Helicopter Parents," *Richmond Alumni Magazine*, Winter 2006, pp. 20–23.

11. Judith R. Shapiro, "Keeping Parents off Campus," *New York Times*, August 22, 2002, p. 23.

12. Liz Marlantes, "Prepping for the Test," *Christian Science Monitor*, November 2, 1999, p. 11.

13. Marlon Manuel, "SAT Prep Game Not a Trivial Pursuit," *The Atlanta Journal-Constitution*, October 8, 2002, p. 1E.

14. Jane Gross, "Paying for a Disability Diagnosis to Gain Time on College Boards," *New York Times*, September 26, 2002, p. A1.

15. Robert Worth, "Ivy League Fever," *New York*

Times, September 24, 2000, Section 14WC, p. 1; Anne Field, "A Guide to Lead You through the College Maze," *BusinessWeek*, March 12, 2001.

16. See the company's Web site, *www.ivywise.com*; Liz Willen, "How to Get Holly into Harvard," *Bloomberg Markets*, September 2003.

17. Cohen quoted in David L. Kirp and Jeffrey T. Holman, "This Little Student Went to Market," *American Prospect*, October 7, 2002, p. 29.

18. Robert Worth, "For $300 an Hour, Advice on Courting Elite Schools," *New York Times*, October 25, 2000, p. B12; Jane Gross, "Right School for 4-Year-Old? Find an Adviser," *New York Times*, May 28, 2003, p. A1.

19. Emily Nelson and Laurie P. Cohen, "Why Jack Grubman Was So Keen to Get His Twins into the Y," *Wall Street Journal*, November 15, 2002, p. A1; Jane Gross, "No Talking Out of Pre-school," *New York Times*, November 15, 2002, p. B1.

20. Constance L. Hays, "For Some Parents, It's Never Too Early for SAT Prep," *New York Times*, December 20, 2004, p. C2; Worth, "For $300 an Hour."

21. Marjorie Coeyman, "Childhood Achievement Test," *Christian Science Monitor*, December 17, 2002, p. 11, citing homework study by University

of Michigan Survey Research Center; Kate
Zernike, "No Time for Napping in Today's Kindergarten," *New York Times,* October 23, 2000,
p. A1; Susan Brenna, "The Littlest Test Takers,"
New York Times Education Life, November 9,
2003, p. 32.

22. See Lawrence H. Diller, *Running on Ritalin: A
Physician Reflects on Children, Society, and Per-
formance in a Pill* (New York: Bantam, 1998);
Lawrence H. Diller, *The Last Normal Child*
(New York: Praeger, 2006); Gardiner Harris,
"Use of Attention-Deficit Drugs Is Found to
Soar among Adults," *New York Times,* September 15, 2005. Ritalin and amphetamine production figures are from Methylphenidate Annual
Production Quota (1990–2005) and Amphetamine Annual Production Quota (1990–2005),
Office of Public Affairs, Drug Enforcement Administration, Department of Justice, Washington, D.C., 2005, cited in Diller, *The Last Normal Child,* pp. 22, 132–133.

23. Susan Okie, "Behavioral Drug Use in Toddlers
Up Sharply," *Washington Post,* February 23,
2000, p. A1, citing study by Julie Magno Zito in
the *Journal of the American Medical Association,*
February 2000. See also Sheryl Gay Stolberg,
"Preschool Meds," *New York Times Magazine,*

November 17, 2002, p. 59; Erica Goode, "Study Finds Jump in Children Taking Psychiatric Drugs," *New York Times*, January 14, 2003, p. A21; Andrew Jacobs, "The Adderall Advantage," *New York Times Education Life*, July 31, 2005, p. 16.

4. THE OLD EUGENICS AND THE NEW

1. See Daniel J. Kevles's fine history of eugenics, *In the Name of Eugenics* (Cambridge, MA: Harvard University Press, 1995), pp. 3–19.
2. Francis Galton, *Hereditary Genius: An Inquiry into Its Laws and Consequences* (London: Macmillan, 1869), p. 1, quoted in Kevles, *In the Name of Eugenics*, p. 4.
3. Francis Galton, *Essays in Eugenics* (London: Eugenics Education Society, 1909), p. 42.
4. Charles B. Davenport, *Heredity in Relation to Eugenics* (New York: Henry Holt & Company, 1911; New York: Arno Press, 1972), p. 271, quoted in Edwin Black, *War against the Weak* (New York: Four Walls Eight Windows, 2003), p. 45; see also Kevles, *In the Name of Eugenics*, pp. 41–56.
5. Letter, Theodore Roosevelt to Charles B. Davenport, January 3, 1913, quoted in Black, *War*

against the Weak, p. 99; see generally Black, *War against the Weak*, pp. 93–105, and Kevles, *In the Name of Eugenics*, pp. 85–95.

6. Margaret Sanger quoted in Kevles, *In the Name of Eugenics*, p. 90; see also Black, *War against the Weak*, pp. 125–144.

7. Kevles, *In the Name of Eugenics*, pp. 61–63, 89.

8. Ibid., pp. 100, 107–112; Black, *War against the Weak*, pp. 117–123; *Buck v. Bell*, 274 U.S. (1927).

9. Adolf Hitler, *Mein Kampf*, trans. Ralph Manheim (Boston: Houghton Mifflin, 1943), vol. 1, chap. 10, p. 255, quoted in Black, *War against the Weak*, p. 274.

10. Black, *War against the Weak*, pp. 300–302.

11. Kevles, *In the Name of Eugenics*, p. 169; Black, *War against the Weak*, p. 400.

12. Lee Kuan Yew, "Talent for the Future," speech delivered at National Day Rally, August 14, 1983, quoted in Saw Swee-Hock, *Population Policies and Programmes in Singapore* (Singapore: Institute of South East Asian Studies, 2005), pp. 243–249 (Appendix A), reprinted at *www.yayapapayaz .com/ringisei/2006/07/11/ndr1983/*.

13. C. K. Chan, "Eugenics on the Rise: A Report from Singapore," in Ruth F. Chadwick, ed., *Ethics, Reproduction, and Genetic Control* (London: Routledge, 1994), pp. 164–171. See also Dan

Murphy, "Need a Mate? In Singapore, Ask the Government," *Christian Science Monitor*, July 26, 2002, p. 1.

14. Sara Webb, "Pushing for Babies: Singapore Fights Fertility Decline," Reuters, April 26, 2006, at *http://www.singapore-window.org/*.

15. Mark Henderson, "Let's Cure Stupidity, Says DNA Pioneer," *Times* (London), February 23, 2003, p. 13.

16. Steve Boggan, "Nobel Winner Backs Abortion 'For Any Reason,'" *Independent* (London), February 17, 1997, p. 7.

17. Gina Kolata, "$50,000 Offered to Tall, Smart Egg Donor," *New York Times*, March 3, 1999, p. A10; Carey Goldberg, "On Web, Models Auction Their Eggs to Bidders for Beautiful Children," *New York Times*, October 23, 1999, p. A11; Carey Goldberg, "Egg Auction on Internet Is Drawing High Scrutiny," *New York Times*, October 28, 1999, p. A26.

18. Graham quoted in David Plotz, "The Better Baby Business," *Slate*, March 13, 2001, at *http://www.slate.com/id/102374/*.

19. David Plotz, "The Myths of the Nobel Sperm Bank," *Slate*, February 23, 2001, at *http://www.slate.com/id/101318/*; and Plotz, "The Better

Baby Business." See also Kevles, *In the Name of Eugenics*, pp. 262–263.

20. I am indebted here to the valuable account of Cryobank in David Plotz, "The Rise of the Smart Sperm Shopper," *Slate*, April 20, 2001, at *http://www.slate.com/id/104633/*.

21. Rothman quoted in Plotz, "The Rise of the Smart Sperm Shopper." For sperm donor qualifications and compensation, see the Cryobank Web site, at *http://www.cryobank.com/index.cfm?page=35*. See also Sally Jacobs, "Wanted: Smart Sperm," *Boston Globe*, September 12, 1993, p. 1.

22. Nicholas Agar, "Liberal Eugenics," *Public Affairs Quarterly* 12, no. 2 (April 1998): 137. Reprinted in Helga Kuhse and Peter Singer, eds., *Bioethics: An Anthology* (Blackwell, 1999), p. 171.

23. Allen Buchanan et al., *From Chance to Choice: Genetics and Justice* (Cambridge: Cambridge University Press, 2000), pp. 27–60, 156–191, 304–345.

24. Ronald Dworkin, "Playing God: Genes, Clones, and Luck," in Ronald Dworkin, *Sovereign Virtue* (Cambridge, MA: Harvard University Press, 2000), p. 452.

25. Robert Nozick, *Anarchy, State, and Utopia* (New York: Basic Books, 1974), p. 315.

26. John Rawls, *A Theory of Justice* (Cambridge, MA: Harvard University Press, 1971), pp. 107–108.

27. I am indebted to David Grewal for illuminating discussion on this point.

28. The phrase comes from Joel Feinberg, "The Child's Right to an Open Future," in W. Aiken and H. LaFollette, eds., *Whose Child? Children's Rights, Parental Authority, and State Power* (Totowa, NJ: Rowman and Littlefield, 1980). It is invoked in connection with liberal eugenics in Buchanan et al., *From Chance to Choice*, pp. 170–176.

29. Buchanan et al., *From Chance to Choice*, p. 174.

30. Dworkin, "Playing God: Genes, Clones, and Luck," p. 452.

31. Jürgen Habermas, *The Future of Human Nature* (Oxford: Polity Press, 2003), pp. vii, 2.

32. Ibid., p. 79.

33. Ibid., p. 23.

34. Ibid., pp. 64–65.

35. Ibid., pp. 58–59. Arendt's discussion of natality and human action can be found in Hannah Arendt, *The Human Condition* (Chicago: Uni-

versity of Chicago Press, 1958), pp. 8–9, 177–178, 247.

36. Ibid., p. 75.

37. The idea that dependence on an impersonal force is less inimical to freedom than dependence on another person finds its parallel in Jean-Jacques Rousseau's social contract: "In giving himself to all, each person gives himself to no one." See Rousseau, *On the Social Contract* (1762), ed. and trans. Donald A. Cress (Indianapolis: Hackett Publishing Co., 1983), Book I, chap. VI, p. 24.

5. Mastery and Gift

1. Tom Verducci, "Getting Amped. Popping Amphetamines or Other Stimulants Is Part of Many Players' Pregame Routine," *Sports Illustrated*, June 3, 2002, p. 38.

2. See Amy Harmon, "The Problem with an Almost-Perfect Genetic World," *New York Times*, November 20, 2005; Amy Harmon, "Burden of Knowledge: Tracking Prenatal Health," *New York Times*, June 20, 2004; Elizabeth Weil, "A Wrongful Birth?" *New York Times*, March 12, 2006. On the moral complexities of prenatal

testing generally, see Erik Parens and Adrienne Asch, eds., *Prenatal Testing and Disability Rights* (Washington, DC: Georgetown University Press, 2000).

3. See Laurie McGinley, "Senate Approves Bill Banning Bias Based on Genetics," *Wall Street Journal,* October 15, 2003, p. D11.

4. See John Rawls, *A Theory of Justice* (Cambridge, MA: Harvard University Press, 1971), pp. 72–75, 102–105.

5. This challenge to my argument has been posed, from different points of view, by Carson Strong, in "Lost in Translation," *American Journal of Bioethics* 5 (May-June 2005): 29–31, and by Robert P. George, in discussion at a meeting of the President's Council on Bioethics, December 12, 2002 (transcript at *http://www.bioethics.gov/transcripts/dec02/session4.html*).

6. For illuminating discussion of the way modern self-understandings draw in complex ways on unacknowledged moral sources, see Charles Taylor, *Sources of the Self* (Cambridge, MA: Harvard University Press, 1989).

7. See Frances M. Kamm, "Is There a Problem with Enhancement?" *American Journal of Bioethics* 5 (May-June 2005): 1–10. Kamm, in a

thoughtful critique of an earlier version of my argument, construes what I call the "drive" or "disposition" to mastery as a desire or motive of individual agents, and argues that acting on such a desire would not render enhancement impermissible.

8. I am indebted to the discussion of this point by Patrick Andrew Thronson in his undergraduate honors thesis, "Enhancement and Reflection: Korsgaard, Heidegger, and the Foundations of Ethical Discourse," Harvard University, December 3, 2004; see also Jason Robert Scott, "Human Dispossession and Human Enhancement," *American Journal of Bioethics* 5 (May-June 2005): 27–28.

9. See Isaiah Berlin, "John Stuart Mill and the Ends of Life," in Berlin, *Four Essays on Liberty* (London: Oxford University Press, 1969), p. 193, quoting Kant: "Out of the crooked timber of humanity no straight thing was ever made."

10. Robert L. Sinsheimer, "The Prospect of Designed Genetic Change," *Engineering and Science Magazine*, April 1969 (California Institute of Technology). Reprinted in Ruth F. Chadwick, ed., *Ethics, Reproduction and Genetic Control* (London: Routledge, 1994), pp. 144–145.

11. Ibid., p. 145.

12. Ibid., pp. 145–146.

EPILOGUE: EMBRYO ETHICS

1. "President Discusses Stem Cell Research Policy," Office of the Press Secretary, the White House, July 19, 2006, at *http://www.whitehouse.gov/news/releases/2006/07/20060719-3.html*; George W. Bush, "Message to the House of Representatives," Office of the Press Secretary, the White House, July 19, 2006, at *http://www.whitehouse.gov/news/releases/2006/07/20060719-5.html*.

2. Press briefing by Tony Snow, Office of the Press Secretary, the White House, July 18, 2006, at *http://www.whitehouse.gov/news/releases/2006/07/20060718.html*; press briefing by Tony Snow, Office of the Press Secretary, the White House, July 24, 2006, at *http://www.whitehouse.gov/news/releases/2006/07/20060724-4.html*; Peter Baker, "White House Softens Tone on Embryo Use," *Washington Post*, July 25, 2006, p. A7.

3. The British legislation, the Human Reproductive Cloning Act 2001, may be found at *http://www.opsi.gov.uk/acts/acts2001/20010023.htm*.

4. Senator Sam Brownback, testimony before Sen-

ate Appropriations Labor, HHS, and Education Subcommittee, Washington, DC, April 26, 2000, quoted in Brownback press release, "Brownback Opposes Embryonic Stem Cell Research at Hearing Today," April 26, 2000, available at *http://brownback.senate.gov/pressapp/record.cfm?id=176080&&year=2000&*.

5. Brownback address at the annual March for Life gathering in Washington, DC, January 22, 2002, quoted in Brownback press release, "Brownback Speaks at Right to Life March," January 22, 2002, available at *http://brownback.senate.gov/pressapp/record.cfm?id=180278&&year=2002&*.

6. My discussion in this section draws on and elaborates the argument I presented in Sandel, "The Anti-Cloning Conundrum," *New York Times*, May 28, 2002, and in my personal statement in *Human Cloning and Human Dignity: Report of the President's Council on Bioethics* (New York: PublicAffairs, 2002), pp. 343–347.

7. Senator Bill Frist, *Congressional Record—Senate*, 107th Cong., 2nd sess., Vol. 148, no. 37, April 9, 2002, pp. 2384–2385; Bill Frist, "Not Ready for Human Cloning," *Washington Post*, April 11, 2002, p. A29; Bill Frist, "Meeting Stem Cells' Promise—Ethically," *Washington Post*, July 18, 2006; Mitt Romney, "The Problem with the

Stem Cell Bill," *Boston Globe*, March 6, 2005, p. D11.

8. Charles Krauthammer, "Crossing Lines," *New Republic*, April 29, 2002, p. 23.

9. For a helpful discussion of the intend/foresee distinction as applied to the cloning and stem cell debates, see William Fitzpatrick, "Surplus Embryos, Nonreproductive Cloning, and the Intend/Foresee Distinction," *Hastings Center Report*, May-June 2003, pp. 29–36.

10. Nicholas Wade, "Clinics Hold More Embryos Than Had Been Thought," *New York Times*, May 9, 2003, p. 24.

11. The phrase "nothing is lost" is from Gene Outka, "The Ethics of Human Stem Cell Research," *Kennedy Institute of Ethics Journal* 12, no. 2 (2002): 175–213. Outka defends the compromise position I criticize. See also the discussion of Outka's "nothing is lost" principle at the President's Council on Bioethics, April 25, 2002, at *http://www.bioethics.gov/transcripts/apr02/apr25session3.html*.

12. In this section and the next, I draw on and elaborate arguments I presented in Sandel, "Embryo Ethics: The Moral Logic of Stem Cell Research," *New England Journal of Medicine* 351 (July 15, 2004): 207–209; and in Sandel, personal

statement, *Human Cloning and Human Dignity.*

13. Paul McHugh, my colleague on the President's Council on Bioethics, advances this view. See "Statement of Dr. McHugh," in the appendix to *Human Cloning and Human Dignity: The Report of the President's Council on Bioethics* (New York: PublicAffairs, 2002), pp. 332–333; and Paul McHugh, "Zygote and 'Clonote': The Ethical Use of Embryonic Stem Cells," *New England Journal of Medicine* 351 (July 15, 2004): 209–211. When McHugh first voiced this suggestion in council discussions, he received criticism bordering on ridicule. But subsequent testimony from Rudolph Jaenisch, an MIT stem cell biologist, offered scientific support for McHugh's distinction between zygote and clonote. See presentation by Rudolph Jaenisch and subsequent discussion, President's Council on Bioethics, July 24, 2003, available at *http://www.bioethics.gov/transcripts/july03/session3.html.*

14. For a critical discussion of this analogy, see Robert P. George and Patrick Lee, "Acorns and Embryos," *New Atlantis* 7 (Fall 2004/Winter 2005): 90–100. Their article responds to Sandel, "Embryo Ethics."

15. I am indebted to Richard Tuck for bringing sorites arguments to my attention, and to David Grewal for pointing out their relevance to the debate about the moral status of embryos.

16. "President Discusses Stem Cell Research Policy," Office of the Press Secretary, the White House, July 19, 2006, available at *http://www.whitehouse.gov/news/releases/2006/07/20060719-3.html.*

17. George J. Annas, "A French Homunculus in a Tennessee Court," *Hastings Center Report* 19 (November 1989): 20–22.

18. In natural procreation, the rate of embryo loss is 60 to 80 percent. According to Dr. John M. Opitz, professor of pediatrics, human genetics, and obstetrics/gynecology at the University of Utah School of Medicine, about 80 percent of fertilized eggs do not survive, and about 60 percent of those that reach the seven-day stage do not survive. See Dr. John M. Opitz, presentation to the President's Council on Bioethics, Washington, DC, January 16, 2003, at *http://www.bioethics.gov/transcripts/jan03/session1.html.* A study published in the *International Journal of Fertility* found that at least 73 percent of natural conceptions do not survive the first six weeks of gestation, and of those that

do, about 10 percent do not survive to term. See
C. E. Boklage, "Survival Probability of Human
Conceptions from Fertilization to Term," *International Journal of Fertility* 35 (March-April
1990): 75–94. For discussion of the ethical implications of embryo loss in natural procreation,
see John Harris, "Stem Cells, Sex, and Procreation," *Cambridge Quarterly of Healthcare Ethics* 12 (2003): 353–371.

Index